Cooperative Extension

AF893049

NRAES–147

Waterborne Pathogens in Agricultural Watersheds

Written by

Barry H. Rosen
Natural Resources Conservation Service
Watershed Science Institute
School of Natural Resources
University of Vermont, Burlington

With contributions by

Richard Croft
Natural Resources Conservation Service
(Retired)

Edward R. Atwill, D.V.M., Ph.D.
School of Veterinary Medicine
University of California–Davis

Susan Stehman, V.M.D.
Senior Extension Veterinarian
Diagnostic Laboratory, Department of
Population Medicine and Diagnostic Science
College of Veterinary Medicine
Cornell University

Susan E. Wade, Ph.D.
Director – Parasitology Section
Diagnostic Laboratory, Department of
Population Medicine and Diagnostic Science
College of Veterinary Medicine
Cornell University

Natural Resource, Agriculture, and Engineering Service (NRAES)

Cooperative Extension • 152 Riley-Robb Hall • Ithaca, New York 14853-5701

NRAES–147
March 2001

Printed by NRAES — Natural Resource, Agriculture, and Engineering Service.
Inquiries invited. (607) 255-7654

First issued in June 2000 by the Watershed Science Institute,
Natural Resources Conservation Service,
United States Department of Agriculture.

ISBN 0-935817-68-9

Library of Congress Cataloging-in-Publication Data

Rosen, Barry H., 1954–
Waterborne pathogens in agricultural watersheds / written by Barry H. Rosen ; with contributions by Richard Croft … [et al.].
p. cm. -- (NRAES ; 147)
Includes bibliographical references (p.).
ISBN 0-935817-68-9 (pbk.)
1. Freshwater microbiology. 2. Agricultural microbiology. I. Title. II. NRAES (Series) ; 147.

QR105.5 .R67 2001
616'.01--dc21

2001030215

Natural Resource, Agriculture, and Engineering Service (NRAES)
Cooperative Extension • 152 Riley-Robb Hall
Ithaca, New York 14853-5701
Phone: (607) 255-7654 • Fax: (607) 254-8770
E-mail: NRAES @CORNELL.EDU • Web site: WWW.NRAES.ORG

Contents

Figures

Tables

Introduction to waterborne pathogens

This technical note provides an introduction to waterborne pathogens, the disease-causing organisms that contaminate water. Key organisms of concern are described in detail, including *Escherichia coli* 0157:H7, *Cryptosporidium parvum* (figure 1), and *Giardia* sp. Indicator bacteria that are normally monitored for water quality are described. Information on viability of organisms in an agricultural setting is presented, along with relevant management practices for controlling waterborne pathogens at their source, thereby reducing the overall pathogen loading within a watershed. The topic of harmful algal blooms is also addressed, although these organisms do not fall neatly into the category of pathogen.

Figure 1 *Cryptosporidium parvum* (courtesy of USEPA, 1999c)

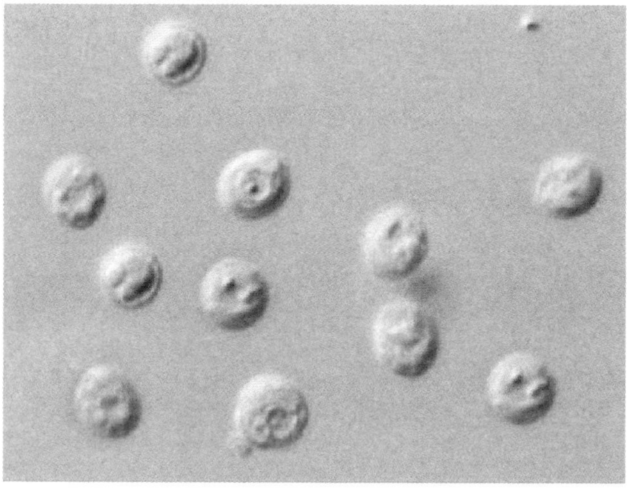

Because the potential does exist for contamination of water with pathogens from agriculture, this technical note represents a proactive approach for reducing this source in watersheds.

Although there are several waterborne pathogens, most disease outbreaks from pathogens are associated with food contamination. Foodborne pathogens are not specifically described in this technical note, although those described may contaminate both food and water. Many other pathogens affect animal health beyond those described in this technical note, and a professional animal health advisor should be consulted for these diseases.

Nature of waterborne pathogens

A pathogen is any agent that causes disease in animals or plants. Pathogens may be a bacterium, protozoan, virus, or worm. *Waterborne zoonotic disease* is a term used to describe illness that is transmitted among animals and humans by water. Transmission from livestock to humans is of particular concern and is emphasized in this technical note.

Most waterborne pathogens are in human and animal feces and enter water along certain pathways. Important pathways include defecation in waterbodies, carried by overland flow and/or subsurface waterflow.

Sources of pathogens

Pathogenic bacteria and protozoa are potentially available from many different animal species in watersheds. Wildlife, pets and companion animals, agricultural animals, and humans are all possible sources of pathogens. In addition, urban development is often associated with an increase in bacteria in runoff (Young and Thackson, 1999).

Human contamination or inadequacies at water treatment plants have been implicated in the previous large-scale waterborne outbreaks while most current waterborne outbreaks are associated with swimming pools and recreational waterbodies (lakes and rivers) (Levy et al., 1998; Upton, 1999).

Agriculture is often attributed as the source of water quality impairments, with both nutrients and bacteria cited as the pollutant. This is partly attributed to *The National Water Quality Inventory Report to Congress*, which is the primary vehicle for informing Congress and the public about general water quality conditions in the United States. This report is compiled by the Environmental Protection Agency (EPA) based on information submitted by the States, tribes, and other jurisdictions in their water quality assessment reports. These reports are issued every even-numbered year and are often referred to as State 305(b) reports, based on the section of the Clean Water Act that describes this reporting mechanism.

The 305(b) reports include information on the degree to which water supports the designated uses, such as recreation or shellfish harvesting. The determination of use support for recreation and shellfish harvesting is based on monitoring for bacterial indicators. The monitoring results for the bacterial indicators are compared against acceptable thresholds, normally the state water quality standards.

The *National Water Quality Inventory: 1996 Report to Congress* lists bacteria as the third leading cause of impairment for rivers and streams, an improvement from being ranked as the leading cause in 1994 (USEPA 1994 305(b) Report). Bacterial impairment of water quality remained second for estuaries. New State and Federal regulations may have economic implications for farmers if stringent source control measures are required for agriculture.

Waterborne pathogens are a particular concern at the watershed scale where they may remain viable and are potentially amplified by agricultural practices. People involved in watershed planning need to understand the basic biological characteristics of these organisms to plan and implement remedial measures, or to avoid practices that could inadvertently promote pathogen viability.

Drinking water

To ensure a safe drinking water supply, treatment normally involves filtration and/or disinfection of the water. Chlorination has long been the preferred method to disinfect drinking water. While effective for bacteria and viruses, it does not kill protozoan cysts (Erlandsen and Meyer, 1984), such as those produced by *Cryptosporidium*.

A collaborative voluntary surveillance system for reporting waterborne-disease outbreaks in humans has been maintained by the Centers for Disease Control (CDC) and the EPA since 1971. The results are used to assess water systems and the type of agents associated with waterborne disease outbreaks, and to improve water quality regulations. The CDC/EPA survey demonstrates an overall improvement in the performance at drinking water facilities (Levy et al., 1998). The survey reported drinking water outbreaks at water treatment plants steadily declined since 1989 (i.e., 72.7% for 1989–1990, 62.5% for 1991–1992, 57.1% for 1993–1994, and 30.0% for 1995–1996).

Recreational water

Exposure to pathogens can occur during swimming or other recreational activities. Exposure is related to ingestion, inhalation, or direct contract with contaminated water. The most recent waterborne outbreaks are associated with recreational waterbodies (lakes and rivers) and swimming pools (Levy et al., 1998; Upton, 1999). Lakes and rivers allow direct exposure of people to water that may contain pathogens. In certain locations these streams and rivers may link to agriculture. The EPA is developing policies to ensure that states and tribes strengthen their water quality standards for recreation (USEPA 1999d).

Principal pathogens of concern

While a large number of viral, protozoal, and bacterial pathogens are potentially shed in human and animal feces, relatively few cause waterborne disease outbreaks. Waterborne infections are often, but not exclusively, spread by water contaminated with intestinal organisms. Free living bacteria and parasites that live in water as their natural environment only sporadically cause skin and respiratory infections in humans.

Table 1 on page 4 lists pathogens causing gastrointestinal illness. This list is from Centers for Disease Control/Environmental Protection Agency (CDC/EPA) report summaries published in the last decade. Pathogens suspended in feces and manure may not survive environmental or treatment conditions long enough to cause infection. Even if these organisms do survive to reach waterbodies, they can be diluted to levels below doses required to infect animals and humans. Because of this dilution effect and water treatment systems used, the vast majority of human infections caused by microbial agents commonly associated with waterborne disease outbreaks are spread by contaminated food or by direct host-to-host contact. For disease outbreaks attributable to waterborne pathogens, the main contributor is water contaminated with human feces (Levy et al., 1998; Upton, 1999) or sewage rather than agricultural operations or wildlife.

Waterborne pathogens of greatest concern have the following characteristics:

- The organisms are shed into the environment in high numbers, or they are highly infectious to humans or animals at low doses.
- The organisms can survive and remain infectious in the environment for long periods, or they are highly resistant to water treatment.
- Some types of bacterial pathogens can multiply outside of a host under favorable environmental conditions.

Table 1 Causes of waterborne disease outbreaks causing gastroenteritis 1989–1996 1/

Type of organism	Agent	Number of outbreaks	Outbreaks associated with drinking water		Outbreaks associated with recreational water	
			surface	ground	natural	pool/ park
Protozoa	*Giardia* sp.	27	12	6	4	5
	Cryptosporidium parvum	21	4	4	2	11
Bacteria with potential for infecting multiple species	*Escherichia coli* O157: H7	11		3	7	1
	Campylobacter jejuni	3	3			
	Salmonella typhimurium	1		1		
	Salmonella java	1				1
	Leptospira grippotyphosa	1			1	
Bacterial infections associated with humans	*Shigella sonnei*	17		7	10	
	Shigella flexneri	2		1	1	
Human viruses	Hepatitis A	3				3
	Norwalk virus	1		1		
	Norwalk-like virus	1				1
	Small round structured virus	1	1			
Acute gastroenteritis	Unidentified cause-many consistent with viral epidemiology	60	8	44	7	1
Other	Cyanobacteria-like bodies	1	1			

1/ Summary of results from studies by Herwaldt et al., 1991; Moore et al., 1993; Kramer et al., 1998; and Levy et al., 1998.

Protozoa
Primary concern

Protozoa of primary concern in watersheds are *Cryptosporidium parvum* and *Giardia* species (abbreviated sp.; figure 2). These protozoa cause mild to severe diarrhea and potentially shorten the lifespan of immune-compromised individuals, such as those with AIDS. These protozoa are found throughout the world, can coexist in the same animal, and are commonly transmitted from the feces to the mouth in animals and humans (i.e., diaper changing). Neither *Cryptosporidium* nor *Giardia* can reproduce outside the host. Illness in humans can be caused by relatively low ingestion levels for both species, with as few as 10 cysts or less. They also produce a waterborne cyst *(Giardia)* or oocyst *(Cryptosporidium)* that is resistant to chemical disinfectants (Sterling, 1990).

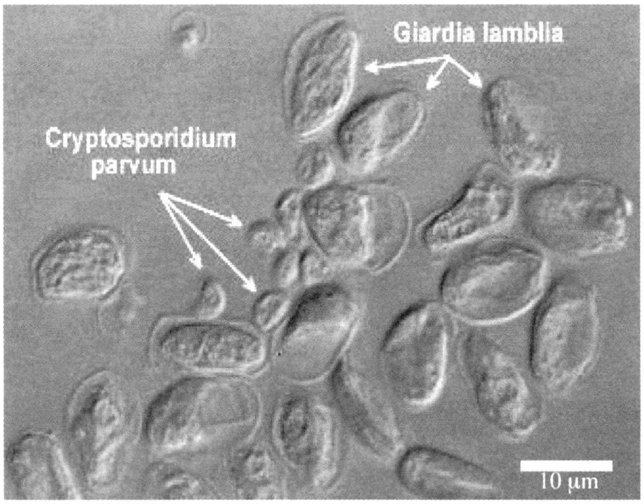

Figure 2 *Giardia* and *Cryptosporidium* micrograph to illustrate relative size

The CDC/EPA survey for 1995–1996 indicated that all of the outbreaks of cryptosporidiosis were associated with recreational water, primarily swimming pools (Levy et al., 1998). Pool-associated outbreaks were caused by chlorine-resistant parasites (e.g., *Cryptosporidium* and to a lesser extent *Giardia*). Prevention is particularly difficult because it requires improved filtration methods as well as education of patrons about hazards associated with fecal accidents, especially in pools frequented by diaper-aged children. This same CDC/EPA survey noted an improvement in the trends from drinking water facilities.

What are protozoa?

Protozoa are microscopic, single-celled organisms that belong to the kingdom Protista. Although they are single-celled, their structure is highly complex. About 66,000 species have been described, of which 10,000 are parasitic. Only a few species are important disease-causing parasites in humans. If they cause disease, they are termed pathogens. While parasites in general obtain food and shelter from their host, they may not harm them.

Parasitic protozoa occur in many different species of animals. Many protozoa are host-specific, but some, such as *Cryptosporidium parvum*, can infect many unrelated host species. Protozoa may complete their lifecycle within one animal or depend on a change of hosts to complete their cycle. Some parasitic protozoa are passed from animal to animal directly, while others form cysts that are resistant to environmental pressures, such as temperature, moisture, and pH, and can be passed from host to host via water. Protozoa are divided into four principal classes according to their mechanism of locomotion (table 2, page 6).

A protozoan lifecycle represents a series of stages through which the organism passes in relation to its environment. In simpler cases an organism reproduces by simple division when food supplies and other conditions are favorable. As conditions change, such as decline of food supply or an increasingly harsh environment, the organism stops dividing and encysts by secreting an outer covering (cyst wall). This cyst wall can be durable. The encysted stage remains until changing conditions (varies by species) induce hatching of the cyst (excystment).

Table 2 Properties of protozoa groups

Class	Description
Mastigophora	Flagellated cells for motility; major pathogens are *Trypanosoma*, *Leishmania*, *Trichomonas*, *Giardia*.
Sarcodina	Amoeba-like motility. Major pathogen is *Entamoeba hystolytica*, the cause of amoebic dysentery.
Sporozoa	All are nonmotile, animal parasites with a complex lifecycle that may require a different host for sexual and asexual reproduction. Do not engulf particulate matter. Major pathogens are *Plasmodium* (malaria), *Toxoplasma* (toxoplasmosis), and *Cryptosporidium*.
Ciliophora	Cells with short, hair-like projections called cilia. Major pathogen is *Balantidium coli*, which causes dysentery in humans.

In wilderness settings, water that appears clean is often contaminated with protozoa, such as *Giardia*. Wildlife contamination is common, and water must be treated before consumption.

Cryptosporidium parvum—*Cryptosporidium parvum* is a protozoan parasite that infects many humans, agricultural livestock (cattle, sheep, goats, pigs, and horses), companion animals (i.e., dogs and cats), and wildlife species, such as mice, voles, and raccoons (Fayer and Ungar, 1986; Fayer, 1997).

Figure 3 *Cryptosporidium* sporozoites (arrows) released from an oocyst-excystation (courtesy of Dr. Timothy Flanigan, Brown University)

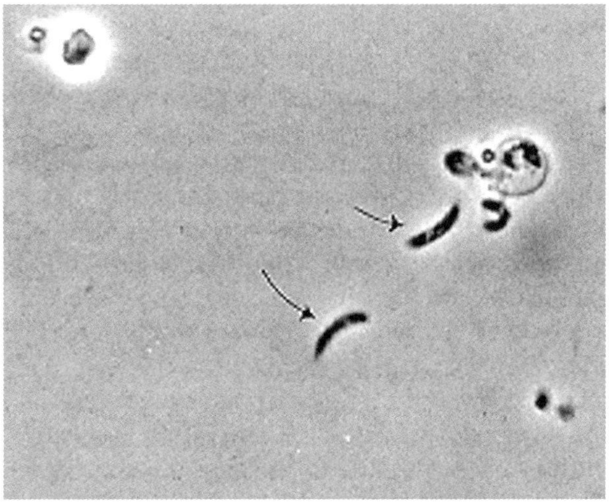

The lifecycle of *Cryptosporidium parvum* may be completed in 1 to 8 days, but varies with the species. The small, colorless, ovoid to spherical oocyst, containing four sporozoites (figure 3), is shed in the feces and is immediately infective to a susceptible animal. When the oocyst is ingested, the sporozoites are released and parasitize the lining cells of the small intestine.

Two kinds of oocysts—thick walled and thin walled—are formed during the sexual phase of the lifecycle. Thick walled oocysts (80% of those formed) are environmentally resistant and are shed in the feces. The thin walled oocysts (20% of those formed) rupture in the intestine, and sporozoites invade new intestinal cells, giving rise to autoinfection. This is one reason that only a few ingested oocysts can ultimately lead to a heavy infection. Oocysts can be shed in feces for several days by an infected animal, with up to 10 million oocysts per gram (about 280 million per ounce) of feces during the peak of shedding (Blewett, 1989; Goodgame et al., 1993; Xiao and Herd, 1994a).

Cryptosporidium parvum was described by Tyzzer (1912) from the small intestine of mice. This species is the one found in most mammals.

It causes illness in young ruminants, dogs, horses, nonhuman primates, and humans. Illness occurs in adult humans, which is unusual among mammals. One strain of *Cryptosporidium parvum* identified is only passed among humans (Pieniazek et al., 1999).

Waterborne outbreaks of cryptosporidiosis have been reported in the United States since 1983. The Internet website **http://www.ksu.edu/parasitology/water** lists 69 *Cryptosporidium* outbreaks through September 1999. The data clearly indicate that sewage contamination and drinking water treatment deficiencies are the major causes of the large-scale outbreaks. Farm animal contamination of water was only reported in three cases.

Despite using state-of-the-art technology, *Cryptosporidium* still breaks through best in water treatment systems (Hayes et al., 1989; Goldstein et al., 1996). The best way to reduce public exposure to *Cryptosporidium* is by optimizing the water treatment process, increasing testing of intake water for the *Cryptosporidium* as well as the treated water, and working closely with public health organizations to analyze risk and quickly identify outbreaks (Mayo Clinic Health Oasis, 1999).

Studies have shown that *Cryptosporidium* sp. oocysts were present in 39 to 87 percent of surface water (rivers, lakes) tested throughout the United States from 1988 to 1993 (Rose et al., 1991; LeChevallier et al., 1991; LeChevallier and Norton, 1995). The source(s) of the organism is typically unknown (Juranek, 1995; Atwill, 1996; Sterling, 1990; Smith and Rose, 1990). In addition, unless species are differentiated, the potential to cause human infection is not known.

An outbreak of cryptosporidiosis in Milwaukee in 1993 is the largest waterborne disease outbreak reported in the United States. An estimated 400,000 people were reported ill. High tributary flows into Lake Michigan because of rain and snow runoff may have transported oocysts great distances into the lake from its watershed, and from there to the water plant intake. Although all applicable water quality standards were being met by the water treatment plant, the facility needed significant upgrades to reduce the risk of *Cryptosporidium* in treated water (MacKenzie et al., 1994; Blair, Internet communications). Outbreaks have been traced to swimming pools in which the filters were not working properly. Contamination resulted from defecation by one of the swimmers while in the pool (Porter et al., 1988).

Only two reported waterborne outbreaks have been directly linked to farms. These two outbreaks occurred in Britain and were traced to contamination of the drinking water by cow manure. One case was a result of surface runoff after a heavy rainfall, and the other was caused by spreading of manure on the watershed of a reservoir (Badenoch, 1990). One study found that after 33 days in river water, an estimated 40 to 45 percent of the oocysts were apparently dead, and after 6 months, 89 to 99 percent were dead (Robertson et al., 1992).

Although severe environmental pressures against oocysts remaining infective when excreted on land exist, only a few oocysts need to remain alive to pose a risk to humans. Robertson et al. (1992) postulated that cow feces might provide additional protection by coating the oocysts, making them less vulnerable to environmental factors.

Experimental studies in healthy humans determined that the infectious dose at which 50 percent of subjects acquired infection was 132 bovine-derived oocysts (DuPont et al., 1995). As few as 30 oocysts were shown to induce cryptosporidiosis (DuPont et al., 1995), but since a dose of less than 30 oocysts was not fed to the volunteers, the minimal infectious dose may be lower.

Production of oocysts generally is limited to livestock that are less than 30 days old (Atwill et al., 1999; Atwill et al., 1998; Xiao and Herd, 1994a; Xiao and Herd, 1994b; Sanford, 1987; Kirkpatrick, 1985) although fecal shedding of oocysts around the time of lambing has been documented in ewes (Xiao et al., 1994). Researchers in Europe have reported evidence of *Cryptosporidium parvum* in adult beef cattle (Scott et al., 1995; Lorenzo et al., 1993), but this has not been confirmed by studies in this country (Wade et al., 1999; Atwill et al., 1998; Xiao and Herd, 1994a; Xiao and Herd, 1994b; Sanford, 1987; Kirkpatrick, 1985).

When livestock feces are deposited on dry land, the transport mechanisms for *Cryptosporidium* oocysts from the manure to a nearby body of water are not well understood. Soil texture, slope, vegetation, and other factors affect the process. Research using intact soil columns and simulated rainfall indicates that *Cryptosporidium parvum* oocysts are carried up to 30 centimeters (about 12 inches) through clay loam, silty loam, and loamy sand. The majority of the oocysts, however, remained within the upper 2 centimeters (about 0.75 inch) of the soil column (Mawdsley et al., 1995). Research has also shown that preferential flow paths formed by soil structure, worms, or roots can move even large particles through the soil (Mawdsley et al., 1995).

One important finding is that the oocysts die when they dry. About 2 hours of dryness was 100 percent lethal for oocysts in the laboratory (Robertson et al., 1992) and were not infective after 1 to 4 days in a barn under summer or winter conditions (Anderson, 1986).

Temperature has also been shown to influence the viability of the oocyst. Oocysts exposed to environmental conditions in a surface soil under cold, but above freezing temperatures, and wet conditions did not lose viability (Ghiorse et al., 1997). On the other hand, freeze-thaw cycles did reduce viability (Ghiorse et al., 1997) and 10 days of temperature below zero degrees Celsius (32 °F) reduced viability by 90 percent. Fayer and Nerad (1996) examined survival in low temperatures and found that *Cryptosporidium parvum* remained viable at minus 10 degrees Celsius (14 °F) and 5 degrees Celsius (41 °F) for 168 hours.

Other species of *Cryptosporidium* occur in other vertebrate groups. Birds are infected by two species: *Cryptosporidium meleagridis*, which is found mostly in the small intestine, and *Cryptosporidium baileyi*, which is found in the respiratory tract and lower intestinal tract. These can cause illness and death, or be asymptomatic in birds, such as chickens, turkeys, ducks, geese, swans, parrots, and finches. Reptiles have at least one species, *Cryptosporidium serpentis*, which can cause clinical illness, chronic gastric disease, weight loss, and death. Upton et al.

(1989) examined 528 individual reptile fecal samples for *Cryptosporidium* sp. oocysts and believe more than one species exists for reptiles. Fish have one species, *Cryptosporidium nasorum*, which kills fish after illness and emaciation. Mammals are not infected by *Cryptosporidium* from birds, reptiles, or fish.

Mammals contract three species: *Cryptosporidium muris*, *Cryptosporidium felis* (Pieniazek et al., 1999) and strains of *Cryptosporidium parvum*. *Cryptosporidium parvum* is the primary organism that causes infection in humans. *Cryptosporidium muris* was described by Tyzzer (1910) from the stomach of mice. It is also found in the stomach lining in cattle from preweaning through adulthood. It is associated with reduced weight gain and milk production. Oocysts may be shed for months. This species has not been reported in humans.

Giardia—*Giardia* is a protozoa frequently found in rivers and lakes that infects the intestinal tract of mammals, such as humans, dogs, cats, bears, muskrats, and beaver, as well as some birds, reptiles, and amphibians (Xiao and Herd, 1994a; Xiao and Herd, 1994b; Xiao et al., 1994; Erlandsen, 1994; Wallis, 1994). Even more widespread than *Cryptosporidium*, most surface water contained *Giardia* cysts (LeChevallier et al., 1991; Wallis et al., 1996). Which of these source(s), including humans, is the most important in surface water has not been clearly demonstrated. The largest outbreak in the 1994–1996 EPA/CDC survey (Levy et al., 1998) came from a community drinking water distribution system (1,449 infected). The source of infection for sporadic cases of waterborne human *Giardia* infection is often not clear (Craun, 1990; Levine et al., 1991; Thompson and Boreham, 1994).

Giardia is a flagellated protozoa with two forms: a motile form (figure 4) or trophozoite, and a cyst. Trophozoite attaches to the surface of the epithelial cells of the upper small intestine near the stomach where it feeds and reproduces. The cyst is ovoid to ellipsoidal and is the environmentally resistant stage that is able to infect a new host. When cysts are ingested by a susceptible host, the lifecycle averages about a week (7–9 days) to be completed (figure 5, page 10). Trophozoites leave the cysts and attach to the surface of the epithelial cells of the small intestine by an adhesive disc. Here they feed and reproduce asexually. The trophozoites form into cysts in the lower part of the small intestine. Cysts leave the host in the feces and may or may not be immediately infective to an animal. Cysts are shed in feces intermittently, often in large numbers. Densities of *Giardia* cysts as high as 400 per liter in sewage effluent and 5 per liter in river water have been reported.

Figure 4 *Giardia* trophozoite (Courtesy of Dr. James M. Smith, McGill University)

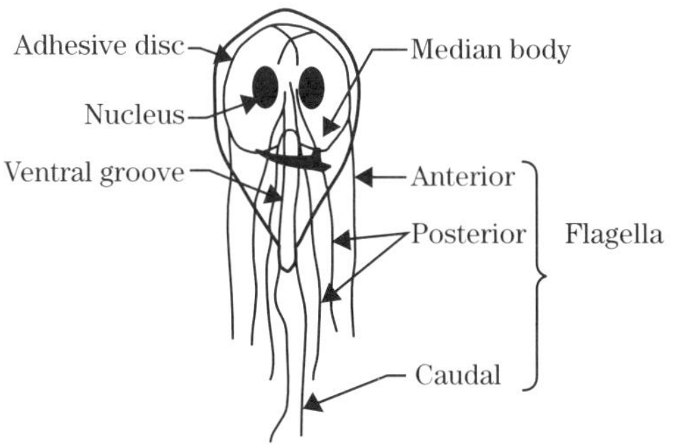

Giardia cysts survive for long periods in the environment. They survive the longest at low water temperatures (0.5 °C or 32.9 °F) (de Regnier et al., 1989; LeChevallier et al., 1991). Most waterborne outbreaks in North America occurred in cooler areas during the spring and fall (Wallis et al., 1996). Cysts are killed by boiling, drying, freeze/thaw cycles, and heating. Management practices that promote any of these measures are effective.

The three recognized species or morphological groups are based primarily on the appearance of structures, called median bodies (Filice, 1952).

Figure 5 *Giardia* lifecycle—binary fission is simply the one cell dividing in two (Courtesy of Dr. James M. Smith, McGill University)

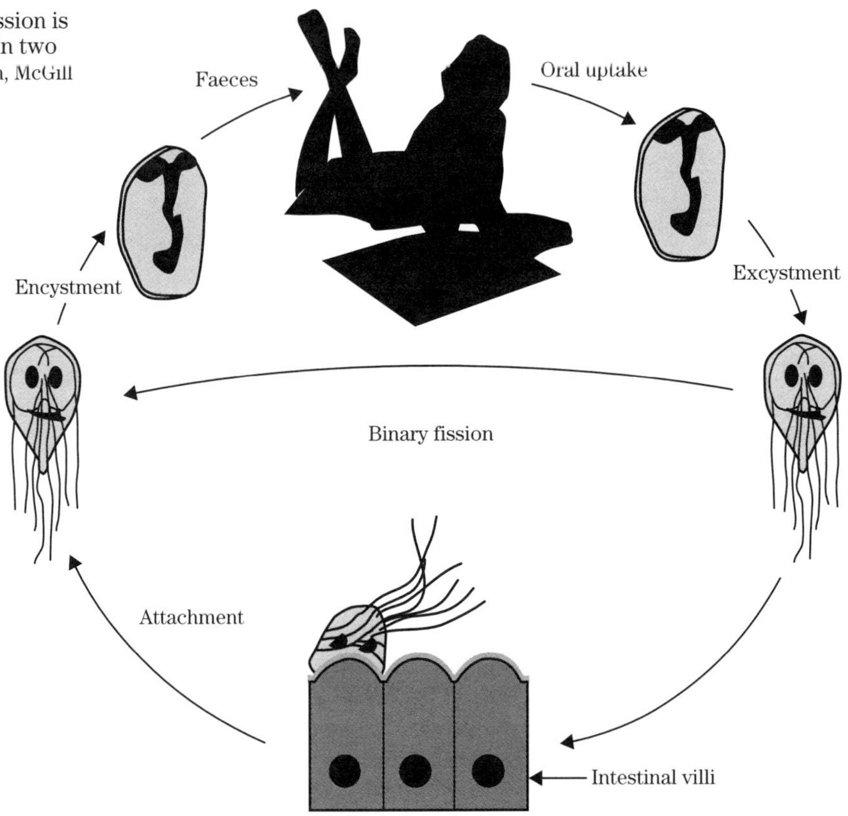

Giardia agilis has a club-shaped median body and is found in amphibians. *Giardia muris* has a round median body and is found in mice and their relatives, birds, and reptiles. *Giardia duodenalis (G. intestinalis* or *G. lamblia)* has a fang-shaped or claw-shaped median body and is found in reptiles, such birds as herons and geese, and in man, cats, dogs, coyotes, ruminants, horses, pigs, and some rodents.

Only the *G. duodenalis* group appears to be composed of distinct species, strains, or races. Since cysts of the *G. duodenalis* type cannot be distinguished morphologically, other methods are needed to determine the source of cysts found in environmental samples. Methods, such as DNA probes and genetic typing, are helping to solve this dilemma. Currently, cyst detection in water samples or in soil and any attempted correlation with infected animals may lead to false conclusions about the origin of the cysts. Until this issue is resolved, the significance of various animals in contaminating the environment, including water supplies, is impossible to determine.

Considerable controversy surrounds the issue of whether *Giardia* sp. cysts obtained from livestock can infect humans (Dupont, 1995). Two reviews (Thompson and Boreham, 1994; Erlandsen, 1994) both concluded that convincing data for the zoonotic potential of *Giardia* still does not exist. Additional review articles debating this issue can be found in an article by Erlandsen (1994). Fayer (1994) also concluded that "no human infection related to these animals has been reported" with respect to pigs, cattle, sheep, and goats.

Finding clear evidence to link livestock to human infection remains difficult. Investigators lack the ability to identify which animal sources *Giardia* sp. cysts came from (Thompson and Boreham, 1994).

Clinical significance of *Giardia* sp. infection in domestic ruminants is currently poorly understood. There are a few reports of diarrhea in calves caused by *Giardia*. The age range of infected cattle is 5-day-old calves to adults, with most of the infected animals being less than 6 months old (Wade et al. 1999). *Giardia* has been considered nonpathogenic in cattle since it is usually found in animals that have normal feces and no sign of disease. Recent studies have indicated that the prevalence of *Giardia* in cattle ranges from less than 14 percent to 100 percent in animals less than 6 months old (Buret et al., 1990; Deshpande and Shastri, 1981; O'Handley et al., 1999; Olson et al., 1997a and b; Pavlasek, 1984; Quilez et al., 1996; Taminelli and Eckert, 1989; Xiao et al., 1993; Wade, et al., 1999).

Protozoa of secondary concern

Waterborne protozoa of secondary importance are those whose transmission to humans does not appear to have a major livestock component or where a waterborne route is rarely documented. Cats serve as the host for *Toxoplasma gondii*. Livestock are not involved in the contamination of water with this protozoan (Acha and Szyfres, 1987; Fayer, 1994). *Balantidium coli*, a ciliated protozoan found in the intestines of humans and pigs, is rarely reported in the United States. Its potential to be transmitted from pigs to humans remains controversial (Acha and Szyfres, 1987). One source of the intestinal amoeba, *Entamoeba histolytica*, is thought to be humans. Livestock have no clear role in human infection (Acha and Szyfres, 1987; Fayer, 1994).

Cyclospora cayetanensis and the opportunistic intestinal organism *Enterocytozoon bieneusi* and *Septata intestinalis* are not known to have a livestock source at this time (Goodgame, 1996).

Bacteria

While this publication focuses on a few bacteria that are harmful to humans, domestic livestock, or both, most bacteria are beneficial for decomposing dead material and releasing nutrients back into the environment for sustenance of plants and animals. As an example, cyanobacteria are a source of oxygen, and their nitrogen-fixing ability is used for rice production in many countries.

Indicators of bacterial contamination

The direct identification and counting of bacteria in water are not practical because the cells are not distinguishable from one another under the microscope. Identification relies on the growth requirements and metabolic functions of cells placed into culture. For water, the procedures for culturing bacteria allow key organisms associated with fecal material to be identified and counted. These key organisms are the indicator bacteria that are used to evaluate the potential for water contamination by pathogens. Although indicator bacteria are not pathogenic in and of themselves, high numbers may indicate fecal contamination from leaky septic tanks, animal manure, or faulty wastewater treatment facilities. Some species also live in soil and on plants and are harmless.

Fecal bacteria have traditionally been used as an indicator of the possible presence of pathogens in surface water and the risk of disease.

Both *E. coli* and enterococci were considered to have a higher degree of association with outbreaks of gastrointestinal illness than fecal coliforms and were recommended by the U.S. Environmental Protection Agency (USEPA, 1986).

What are bacteria?

Bacteria are a group of micro-organisms that lack membrane-bound organelles, and hence are considered simpler than plant and animal cells. They have a cell wall, and some have an outer protective layer. Many antibiotics act by blocking synthesis of the bacterial cell wall. Most bacteria are unicellular and may have various shapes: spherical (coccus), rod-shaped (bacillus), comma-shaped (vibrio), spiral (spirillum), or corkscrew-shaped (spirochete). Generally, they range from 0.5 to 5.0 micrometers. Motile species (those that can move on their own) bear at least one fine hair (flagella) arising from their surface. Many possess an outer, slimy capsule, and some have the ability to produce an encysted or resting form (endospore). Bacteria reproduce asexually by simple division of cells and rarely by conjugation.

Bacteria are so widespread that only the most general statements can be made about their life history and ecology. They are everywhere on Earth, even in the most hostile of environments. Some live in soil, plants, or water; others are parasites of humans, animals, and plants. Many parasitic bacteria do not harm their hosts; some cause diseases by producing toxins.

Bacteria have a wide range of environmental and nutritional requirements. They can be classified into three groups based on their need for oxygen. **Aerobic bacteria** thrive in the presence of oxygen and require it for continued growth and existence. **Anaerobic bacteria** thrive in oxygen-free environments. **Facultative anaerobes** can survive in either environment, although they prefer the presence of oxygen.

Cyanobacteria, commonly called blue-green algae, are a separate group of bacteria that deserves mention here. Cyanobacteria have the ability to fix atmospheric nitrogen into usable organic molecules. They have chlorophyll and are photosynthetic, and although they are not pathogens, they can produce potent toxins. Many regulate their buoyancy, often floating to the surface of a waterbody were livestock have easy accesses to concentrated populations of organisms. Cattle drinking water contaminated with cyanobacterial toxins can die if sufficient toxin is consumed. No human fatalities have been reported. Potential toxins from cyanobacteria have only been a recent concern of water treatment plants. Drinking water facilities are often faced with trihalomethane problems associated with the necessity to chlorinate water that has excess algal growth.

Farm ponds also support a luxuriant growth of cyanobacteria as a result of nutrient enrichment. Excess algae becomes a major problem when cell growth stops and the organisms begin to die. As this happens, decomposing bacteria consume oxygen (respiration) as they breakdown the dead cyanobacteria cells, causing oxygen depletion of the waterbody. This may lead to fish mortality.

Total coliform (table 3) is the broadest category of indicator bacteria and was originally believed to indicate the presence of fecal pollution. Numerous nonfecal sources make this indicator too generic.

Fecal coliform, a subgroup of total coliform, originates from the intestinal tract of warm-blooded animals. This subgroup is the most commonly used indicator of bacterial pollution in watersheds. *Escherichia coli (E. coli)* is a member of the fecal coliform subgroup. This subgroup is used because it correlates well with illness from swimming and can cause gastrointestinal problems.

Fecal streptococci, also called fecal strep, are another grouping of bacteria, similar to the coliforms that are associated with feces from warm-blooded animals. The ratio of fecal coliform to fecal strep may indicate something about the source of the bacteria (table 3).

Enterococci are a subgroup of fecal strep bacteria. This subgroup is used because it correlates well with human illness in recreational waterbodies. Coliphage, a virus that uses coliform bacteria as a host, can also be used as an indicator of fecal contamination.

State and local government agencies commonly monitor for coliform bacteria. Data from monitoring bacterial are reported to the EPA. States and a local health agency may have more stringent standards than the national guidelines. The

Table 3 Comparison of fecal bacteria water quality indicators commonly used (Landry and Wolfe, 1999)

Microbial indicator	Properties	Federal standard [1/] primary contact [2/]
Total coliforms (TC)	• Originally believed to indicate the presence of fecal pollution • Widely distributed in nature: soils, water, flora, fauna • Contains members of *Escherichia*, *Citrobacyter*, *Klebsiella*, and *Enterobacter* identified by incubation at 35 °C	1,000 CFU/100 mL
Fecal coliforms (FC)	• Subgroup of TC • Coliforms that originate specifically from intestinal tracts of warm-blooded animals • Cultured by increasing the incubation temperature to 44.5 °C • Remains the predominant indicator used to assess bacterial pollution in watersheds	200 CFU/100 mL
Escherichia coli	• Member of the FC group • Presence correlates with illness from swimming in both fresh and marine water • Has been shown epidemiologically to cause gastrointestinal symptoms • O157:H7 is a toxin-producing strain of this common bacterium	126 CFU/100 mL
Fecal streptococci (FS)	• Differ from FC: (1) less dominant in feces (2) not known to multiply in the environment (3) die off more rapidly • Found in intestinal tracts of warm-blooded animals and may be found on vegetables and insects • FC/FS ratio used to determine fecal source • Includes *Streptococcus faecalis*, *S. faecium* *S. bovis*, *S. equinus*, and *S. avium*	FC/FS ratio > 4.0 (human source) 0.7 < FC/FS ratio < 4.0 (multiple sources) FC/FS ratio < 0.7 (animal source)
Enterococci	• Subgroup of the FS including *Streptococcus faecalis*, *S. faecium*, and *S. avium* • Commonly found in intestinal tracts of humans and other warm-blooded animals • Presence correlates well with illness from both fresh and marine water	33 CFU/100 mL

1/ Individual states may have stricter standards, but not lower. CFU = colony forming units.
2/ Primary contact water includes recreational use, such as swimming and fishing.

Federal standards are shown in table 3 (page 13). The EPA has targeted 2003 for all states to start using *E. coli* and enterococci as indicators instead of fecal and total coliform (USEPA 1999d). The concept is based on information that these two indicators are better for predicting health risk in primary contract reaction water.

One important point to note on bacterial indicators is that potential pathogens, such as *Cryptosporidium parvum*, *Shigella*, sp. and *E. coli* O157:H7, may be in water that meets all bacterial water quality standards. This clearly indicates that better detection methods are needed for pathogens in water, such as Method 1623 for *Giardia* and *Cryptosporidium* (USEPA, 1999c).

Bacteria of primary concern

Escherichia coli O157:H7 (E. coli O157:H7)—The *E. coli* referred to during general water quality tests (total coliforms, fecal coliforms, *E. coli*) are not necessarily pathogenic strains, but instead are an indicator of general fecal contamination. When used as an indicator of water quality, *E. coli* refers to the harmless strains that help maintain normal intestinal functions (Draser and Hill, 1974).

E. coli O157:H7, however, is a potentially deadly bacteria that can cause bloody diarrhea and dehydration. It is an unusually infectious organism with as few as 10 cells causing illness. Young children, the elderly, and persons with compromised immune systems are the most susceptible. The combination of letters and numbers in the name of the bacterium refers to the specific molecular markers found on its cell surface that distinguish it from other types of *E. coli.* Although most strains are harmless and live in the intestines of healthy humans and animals, the O157:H7 strain produces a powerful toxin capable of causing severe kidney failure and can break down the lining of the intestine in humans. *E. coli* O157:H7 has been isolated from feces of cattle, white-tailed deer, sheep, dogs, horses, birds, and flies.

How one is infected with E. coli O157:H7—The infection is usually acquired by eating food or drinking liquids containing the bacteria. The bacteria in the guts or hides of livestock can contaminate meat during the slaughtering process. Eating undercooked meat, especially undercooked ground meat, is the most common way of getting the infection. Person-to-person transmission can occur if infected persons, especially food handlers, do not wash their hands. Drinking unpasteurized milk or fruit juice, or swimming in or drinking sewage-contaminated water can also cause infection.

The role livestock have with E. coli O157:H7—Although this organism is not pathogenic to cattle themselves, calf water troughs and moist mixed cattle rations have been cited as sources of *E. coli* O157:H7 on farms (Hancock et al., 1997d). *E. coli* O157:H7 has been shown to persist in water trough sediment for at least 4 months and may even grow in the environment (Hancock et al., 1997d). Two surveys of dairy and beef cattle (Hancock et al., 1997a and b) showed wide-spread distribution of *E. coli* O157:H7 in feedlots, ranging from 63 to 75 percent of the 100 herds sampled. However, the data also showed that the entire herd was not affected. Only 1 to 2 percent of the cows in any given herd carried *E. coli* O157:H7.

E. coli O157:H7 appears to be only a transient member of the bacterial flora of a cow, colonizing cattle from 1 to 2 months, rather than being a long-term carrier (Besser et al., 1997). One study showed that *E. coli* O157:H7 increased with dietary stress (Cray et al., 1995). Other influences on the *E. coli* O157:H7 are related to season (found more frequently after May 1) and cow age, more common in cows 3 to18 months old (Hancock et al., 1997c). According to the National Animal Health Monitoring System (Wells, 1999-Internet communications), herd size also has an influence, with *E. coli* O157:H7 more common in larger herds. *E. coli* O157:H7 was not associated with manure application to grazing lands (Hancock, 1997b).

Campylobacter—Several species of *Campylobacter* can cause infection in humans (e.g., *C Campylobacter jejuni*, *Campylobacter coli*, *Campylobacter lardis*, *Campylobacter fetus* subspecies *fetus*). *Campylobacter jejuni* accounts for almost all of the diagnosed cases.

The majority of human campylobacteriosis occurs as sporadic cases as opposed to outbreaks involving large numbers of people (Tauxe, 1992). *Campylobacter jejuni* is common in the environment and is shed in the feces of humans, livestock, and wildlife, including birds. It is also found in a variety of surface water, stream sediment, and sewage effluents (Tauxe, 1992; Stern, 1992; Altekruse et al., 1999). The primary routes of transmission appear to be ingestion of contaminated foods (primarily poultry and raw milk), ingestion of untreated surface water, and contact with pets (primarily dogs) suffering from diarrhea. Direct human-to-human transmission rarely occurs (Tauxe, 1992; Altekruse et al., 1994; Franco and Williams, 1994; Adak et al., 1995).

Cattle and poultry feces and effluent from poultry processing facilities have been shown to contain *Campylobacter jejuni* that in some cases are similar to human isolates (Tauxe, 1992; Stern, 1992; Altekruse et al., 1994; Koenraad et al., 1995).

Campylobacter jejuni can survive for a limited time in stream water (Terzieva and McFeters, 1991). As few as 500 to 800 organisms appear sufficient to cause clinical illness in humans (Robinson, 1981; Black et al., 1988). *Campylobacter* sp. from water can be detected throughout the year, with more isolates found during the cold seasons in temperate climates (Carter et al., 1987; Bolton et al., 1987). *Campylobacter* can survive in a nonculturable state in cold water and still be infectious to livestock (Stern et al., 1994). Unchlorinated drinking water has been identified as a source of infection to cattle herds (Humphrey, 1987).

Despite these findings, documented cases of human campylobacteriosis attributable to water contaminated with livestock manure or livestock effluent are uncommon. Better DNA fingerprinting techniques and more sensitive and specific methods for identifying *Campylobacter jejuni* in water and other environmental sources should help identify the primary source(s) of this common cause of gastroenteritis in humans.

Salmonella—*Salmonella* species cause diarrhea and systemic infections. These infections can be fatal in particularly susceptible persons, such as the immuno-compromised, young children, and the elderly. An estimated 800,000 to 4 million human infections occur each year in the

United States, most of them as individual cases apparently unrelated to outbreaks. The majority of outbreaks are associated with foodborne illness. More than 2,200 reported serotypes of *Salmonella* are found in a variety of host species.

Animals used for food production are common carriers of *Salmonella*. Foods often implicated in outbreaks include poultry and poultry products, meat and meat products, dairy products, egg products, seafood, and fresh produce. *Salmonella enteritidis*, now one of the most commonly isolated serotypes from human cases, is highly associated with ingestion of contaminated shell eggs (Mishu et al., 1994). Recent papers indicate that the home environment may be a much more common source than food.

Whether manure from poultry operations or the effluent from poultry processing facilities is a significant source of waterborne salmonellosis in humans is unclear at this time. One outbreak in a large California poultry flock was associated with the same type of *Salmonella enteritidis* found in creek water and sewage effluent from a nearby treatment plant. This indicates that livestock can be the source or recipient of *Salmonella* infections (Willoughby et al., 1995).

A concise review of the role of livestock in the annual incidence of waterborne human salmonellosis is difficult, at best. Only two outbreaks of waterborne salmonellosis have been reported since 1989. One was associated with wild bird contamination of water in a storage tank, and the other was associated with water exposure in a recreational pool. The most noticeable observation regarding human salmonellosis is the lack of reported waterborne outbreaks definitively traced to livestock (Acha and Szyfres, 1987; Ziprin, 1994).

Salmonella newport has been shown to survive for extended periods in freshwater sediment (Burton et al., 1987). *Salmonella typhimurium* is another common isolate from cattle (Graeber et al., 1995). In Germany, isolates of *Salmonella typhimurium* that were obtained from nearby calf-rearing facilities were not the strains contaminating associated coastal water nor the isolates obtained from nearby human patients (Graeber et al., 1995). Studies such as this provide no assurance regarding the modes of transmission in the United States.

Secondary concern

This group includes waterborne bacteria of secondary importance. In this group, waterborne transmission to humans does not appear to have a major livestock component or a waterborne route is rarely documented.

Yersinia—The estimated number of human infections with *Yersinia* species (yersiniosis) in the United States is 3,000 to 20,000 per year. This is far less than the reported number of campylobacteriosis and salmonellosis cases (Roberts, 1989; Todd, 1989). Most *Yersinia* infections in the United States result from *Yersinia enterocolitica*. One of the primary transmission routes is food (Feng and Weagant, 1994). The proportion of sporadic cases attributable to food versus water versus other routes of transmission is not well established. Water is not considered a major source of infection in North America.

Yersinia species are widespread in water (streams and lakes), foods (pork products, vegetables, dairy products, tofu, seafood), wild and domestic animals, and humans, but many of the strains obtained from these sources were not pathogenic for man. Relative to *Campylobacter jejuni*, *Yersinia enterocolitica* is better able to survive in stream water (Terzieva and McFeters, 1991). Despite swine being considered as one of the primary environmental sources of these bacteria, documented waterborne outbreaks of human yersiniosis have not been definitively linked to livestock (Acha and Szyfres, 1987). Because of the low number of human yersiniosis cases, this pathogen should be classified as a waterborne disease of secondary importance.

Brucella sp.—The waterborne route of transmission is not known to play a significant role in human infection with *Brucella* sp. (Acha and Szyfres, 1987; Chomel et al., 1994).

Leptospirosis interrogans —Human infections with isolates of *Leptospirosis interrogans* are rare. They are confined mostly to direct contact with infected animals, as might occur among veterinarians and slaughterhouse workers (Acha and Szyfres, 1987; Heath and Johnson, 1994).

Although occasional waterborne outbreaks have been reported, the role of livestock is unclear (Shaw, 1992; Jackson et al., 1993). *Leptospirosis* species are carried by many different mammal hosts.

Viruses

Little evidence shows that viruses shed in the excrement of livestock have posed a waterborne threat to human health in the United States (Cliver, 1994). Runoff from spreading of municipal sludge and manure may be a source of viruses to waterbodies. This source is addressed in a later section. In addition, septic tank effluent may be the most significant source of pathogenic viruses in the subsurface environment (CAST, 1992). Interspecies transmission of rotaviruses has been demonstrated experimentally. Livestock-derived rotaviruses are generally not considered a source of human waterborne infection (Acha and Szyfres, 1987).

What are viruses?

Viruses are the smallest known agents of disease that infect plants, animals, and even bacteria. They use their host cells for reproduction and are unable to reproduce outside their host. When viruses are outside host cells, they exist as DNA or RNA surrounded by a protein coat or capsid ranging from 20 to 300 nanometers (1 nanometer = one billionth of an inch). No other structures are typically found in cells of other organisms, such as a nucleus or chloroplast. Because a virus does not have any of these organelles, it has no metabolism.

Viruses are not active unless they are within living cells. When it comes in contact with a host cell, the virus can inject its genetic material into the cell, leaving its protein shell behind. Once within the host cell, the virus uses the cell's own cellular processes, such as protein synthesis, to produce more viruses. In essence, the virus forces the cell to replicate the virus' own genetic material and protective shell. Once replicated, the new viruses leave the host cell and are ready to invade others. Some viruses may remain dormant inside their host cell (lysogenic phase) for long periods, causing no obvious change in the host cell. However, when a dormant virus is stimulated (lytic phase), new viruses are formed and burst out of the host cell, killing it and going on to infect other cells.

Hundreds of known viruses cause a wide range of diseases in humans, other animals, insects, bacteria, and plants. Within a species, 100 or more viruses may infect that species alone. Most viruses are specific to a single host species. A few are more general and are capable of infecting one or more animal species, including humans.

Groups of viruses are classified by the arrangement and type of their genetic material (table 4). The animal virus group is subdivided into the following subgroups: double-stranded DNA, single-stranded DNA, double-stranded RNA, single-stranded RNA, and retroviruses, a unique kind of single-stranded RNA virus.

Table 4 Partial list of types of human virus associated with feces 1/

Group	Type 2/	Water relationship
Adenovirus	DS DNA	Isolated from sewage, rivers, lakes, ground water, drinking water, and recreational bathing water (swimming pools are a major source)
Astrovirus	SS RNA	Feces, therefore, inferred in sewage
Caliciviruses	SS RNA	Feces, therefore, inferred in sewage
Hepatitis	SS RNA	Sewage and polluted rivers
Norwalk-like	SS RNA	Municipal drinking water contaminated with sewage, recreational bathing
Rotavirus	DS RNA	Sewage, rivers and lakes, estuarine and marine water
Coxsackievirus	SS RNA	Feces, therefore, inferred in sewage

1/ Adapted from AWWARF Workshop on Emerging Pathogens, 1999.
2/ DS = double-stranded, SS = single-stranded.

Harmful algal blooms

Algae are a natural part of marine and freshwater environments. They are the base of the food web and provide oxygen to the water. Some harmful algal blooms, like toxic *Pfiesteria* outbreaks, cannot be visibly detected and can be detrimental even at low concentrations. In other cases, like certain red and brown tides, harmful effects occur when the algae reach concentrations high enough to discolor the water. However, not all algal blooms that discolor the water are harmful. Many red tides have no negative effects on marine life, people, or the environment.

Some kinds of algal blooms, like some kinds of red tides, are harmful because the algac produce one or more toxins that poison fish or shellfish. They also can pose human health risks when people come in contact with affected water. These toxic algal blooms may also kill seabirds and other animals indirectly when the toxins are passed up the food chain. Certain kinds of these toxic algal blooms can cause human health problems via contaminated seafood. Ciguatara fish poisoning, amnesic shellfish poisoning, and paralytic shellfish poisoning are examples. However, there is no evidence that *Pfiesteria*-related illnesses are associated with eating fish or shellfish. Most algal blooms are not toxic, but they are still considered harmful if they reduce the amount of light or oxygen in the water, consequently killing sea grasses, fish, or other marine life.

Pfiesteria

Recent attention has be focused on *Pfiesteria piscicida*, which has been associated with fish lesions and fish kills in coastal water from Delaware to Alabama (USEPA, 1999a, Internet communication). These organisms are believed to be native, not introduced species, and are probably common inhabitants of estuarine water within their range. These microbes have not been found in freshwater lakes, streams, or other inland waterbodies.

Figure 6 *Pfiesteria* scanning electron micrograph (Courtesy of Dr. Karen Steidinger, Florida Marine Research Institute)

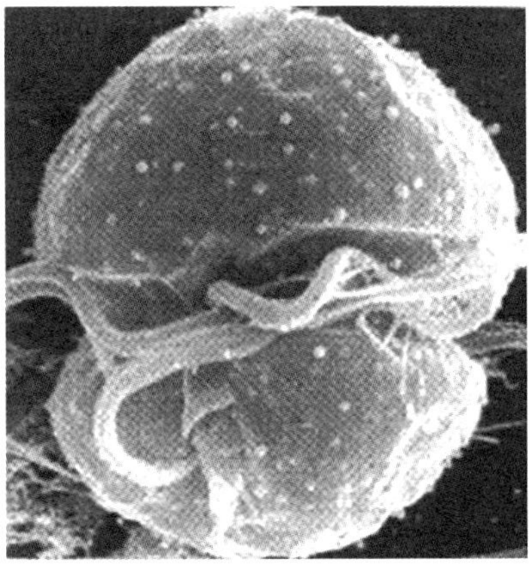

Pfiesteria identification—*Pfiesteria* belongs to the dinoflagellates group of algae. These algae are microscopic, normally free-swimming, single-celled organisms (figure 6).

Although many dinoflagellates are plant-like and obtain energy by photosynthesis, others, including *Pfiesteria*, are more animal-like and acquire some or all of their energy by eating other algae and often incorporating their prey's chloroplast into their own cells. The vast majority of dinoflagellates are not toxic. *Pfiesteria*, however, is a known toxin producing dinoflagellate.

Discovered in 1988 by researchers at North Carolina State University, *Pfiesteria piscicida* is now known to have a highly complex lifecycle with 24 reported forms, a few of which can produce toxins. *Pfiesteria* was named in honor of the late Dr. Lois Pfiester, who contributed much of what we know today about the complex lifecycles of dinoflagellates.

A few other potentially toxic dinoflagellate species with characteristics similar to *Pfiesteria* have been identified, such as *Cryptoperidiniopsis* (Steidinger et al., *in prep*), which also is a species that feeds on microalgal prey. In North Carolina and Maryland, this species can occur in brackish systems with *Pfiesteria piscicida*. These newly-identified species are referred to as *Pfiesteria*-like organisms, and they occur from Delaware to the Gulf of Mexico.

***Pfiesteria* and human health problems**—Preliminary evidence suggests that exposure to water where toxic forms of *Pfiesteria* are active may cause memory loss, confusion, and a variety of other symptoms including respiratory, skin, and gastro-intestinal problems. It has been shown that similar human health effects can be caused by exposure to *Pfiesteria* toxins in a laboratory setting. To date, other *Pfiesteria*-like organisms have not been shown to cause human illness.

Pfiesteria is not contagious or infectious and cannot be caught like a cold or flu. No evidence shows that *Pfiesteria*-related illnesses are associated with the *consumption* of finfish, shellfish, or crustaceans, such as crabs, lobsters, and shrimp. Any human health problems associated with the microbe stem from its release of toxins into river and estuarine water and human contact with that water, rather than the organism infecting a person.

The Centers for Disease Control and Prevention, in cooperation with state health departments in Delaware, Florida, Georgia, Maryland, North Carolina, South Carolina, and Virginia, have established a surveillance system to collect reports of human illness thought to be related to exposure to *Pfiesteria* and *Pfiesteria*-like organisms in estuarine water. This and other ongoing research efforts are expected to further delineate the nature, extent, and duration of any *Pfiesteria*-related human health effects.

Nutrients and Pfiesteria—Nutrients, such as nitrogen and phosphorus, are thought to encourage the growth of *Pfiesteria* populations by stimulating the growth of algae that *Pfiesteria* feeds on when in its nontoxic forms. Some evidence suggests that nutrients may also directly stimulate the growth of *Pfiesteria*, but more research is needed to show this conclusively. At this time the precise role that nutrients and other factors may play in promoting toxic outbreaks of *Pfiesteria* is not clear and is an area of active research.

Excess nutrients are common pollutants in coastal water. Chief sources of nutrient pollution in coastal areas are sewage treatment plants, septic tanks, polluted runoff from suburban landscapes and agricultural operations, and air pollutants that settle on the land and water.

Causes of toxic Pfiesteria outbreaks—The exact conditions that cause toxic outbreaks of *Pfiesteria* to develop are not fully understood. Scientists generally agree that a high density of fish must be present to trigger the shift of *Pfiesteria* cells into toxic forms. However, other factors may contribute to toxic *Pfiesteria* outbreaks by promoting the growth of *Pfiesteria* populations in coastal water. These factors include warm, brackish, poorly flushed water and high levels of nutrients.

What is being done about Pfiesteria—State and Federal agencies are working closely with local governments and academic institutions to address the problems posed by *Pfiesteria*. Federal agencies involved in the effort include the U.S. Environmental Protection Agency, the National Oceanic and Atmospheric Administration, the Centers for Disease Control and Prevention, the National Institute of Environmental Health Sciences, the Food and Drug Administration, the U.S. Geological Survey, and the U.S. Department of Agriculture. Together with state departments of health and natural resources, these agencies are working to:

- manage the risk of human health effects by monitoring and rapid response through river closures and public health advisories
- direct funding for technical expertise to do *Pfiesteria*-related research and monitoring
- make current and accurate information widely available to public
- understand and address the causes of *Pfiesteria* outbreaks, especially the possible role of excess nutrients

Helminths

Helminths are worms that may be free-living or parasitic in plants and animals. The parasitic worms of greatest concern in water are Platyhelminthes or flatworms (flukes and tapeworms) and Nematoda (roundworms). Most flukes and tapeworms require a minimum of two hosts to complete their lifecycle.

Infection with one or a few *Ascaris* sp. may not be noticed. Infection with numerous worms may result in pneumonia during the migratory phase when larvae that have hatched from the ingested eggs penetrate into the lungs. Vague digestive tract discomfort sometimes accompanies intestinal infection, but small children that have more than a few worms may have intestinal blockage because of the worms' large size. No worms stay on the path that is optimal for their development. Those that wander can penetrate into tissues and locate in various organ systems of the human body, causing complications.

Ascaris lumbricoides is one of the largest parasitic roundworms and is the most common parasite found in humans. It is estimated that 20 to 25 percent of the world's population is infected with this nematode (World Health Organization, 1992). The adult female of this species can measure up to 18 inches long (males are generally shorter). The adult worms live in the small intestine and eggs are passed in the feces. A single female can produce up to 200,000 eggs each day.

Ascarus suum, a parasite common in pigs, has larvae that will migrate to the lungs and die. This can cause a particularly serious form of pneumonia. Adult worms of this species develop in small children's intestines.

Nematode eggs, such as those of *Ascaris* sp., can contaminate crops when irrigation water has not been adequately disinfected. Sludge may also be a source of crop contamination if inadequate pretreatment is used. Humans can be infected if they eat raw produce that is contaminated with live ascaris eggs.

Hookworm is a nematode that is endemic in moist tropical and subtropical regions. When inadequately treated sewage is used on croplands, in combination with the naturally high soil moisture, one can expect hookworm infection. The infection is usually contracted by persons walking barefoot over contaminated soil.

Pathogens in the environment

Sources of waterborne pathogens

Water plays a major role in mobilizing and transporting micro-organisms. Rainfall washes organisms from feces or vegetation surfaces and directs them into the soil or along the land surface and thus into surface water. Water contaminated by pathogens is often a health risk for swimming and potentially hazardous for other recreational uses. It must receive extensive treatment for drinking water use. A major challenge in watersheds is to limit the number of micro-organisms reaching surface and ground water, thereby avoiding significant contamination. The goal is to minimize the number of pathogens to a level lower than that needed to cause infection, thereby protecting water users.

Several potential sources of pathogens are in the environment. Complex pathways for their distribution are common. Only three waterborne disease outbreaks have been associated with agriculture. The vast majority are associated with pools, sewage contamination associated with a drinking water facility or inadequacies or failures at these facilities, or biofilms in post-treatment distribution systems. In contrast, agriculture continues to be recognized as a major cause of not attaining water quality based on indicator bacteria standards (USEPA, 1996).

Although sewage and septic systems are the main source of contamination, identifying the potential impacts from agricultural sources is important. Some of the potential pathways for pathogens in the environment are illustrated in figure 7.

For indicator bacteria, agriculture continues to be recognized as a major cause of not attaining water quality standards (USEPA, 1996).

Domestic livestock

Animal feeding operations and grazing lands are all potential sources of pathogens. How livestock are kept varies, but the trend in modern production agriculture is to confine animals so careful attention can be given to maximizing meat, fiber, or dairy yields. These are known as animal feeding operations (AFOs) or concentrated animal feeding operations (CAFOs), based on size and discharge characteristics, and are described in the EPA-USDA Unified National Strategy for Animal Feeding Operations (USEPA, 1999b, based on 40 CFR 122.23(b) and app. B). Most poultry, swine, and increasingly, dairy cattle, are confined, creating a waste stream that may have multiple stages for manure treatment and disposal. Beef cattle, sheep, and, to a lesser extent, dairy animals and swine may graze on rangeland or pastureland during part of this production period.

Bare areas, such as open lots with heavy animal traffic, have the greatest potential for pathogen (and nutrient) runoff into surface water or leaching into ground water. Direct deposit into streams is also an obvious source. Onfarm recycling of manure has the potential to carry pathogens into surface water if proper application practices designed to prevent runoff are not followed.

Land application of sewage and sludge

Biosolids (sewage sludge) are aptly described as a concentrate of pathogens and must require prudent treatment, handling, and land application and crop utilization techniques to avoid infection of humans and animals. The class of sludge directly affects pathogen load. Class A sludge is treated, usually through high temperature, with the goal of reducing pathogens below detection levels. This sludge can be used without

Figure 7 Potential pathways for waterborne pathogen movement in watersheds

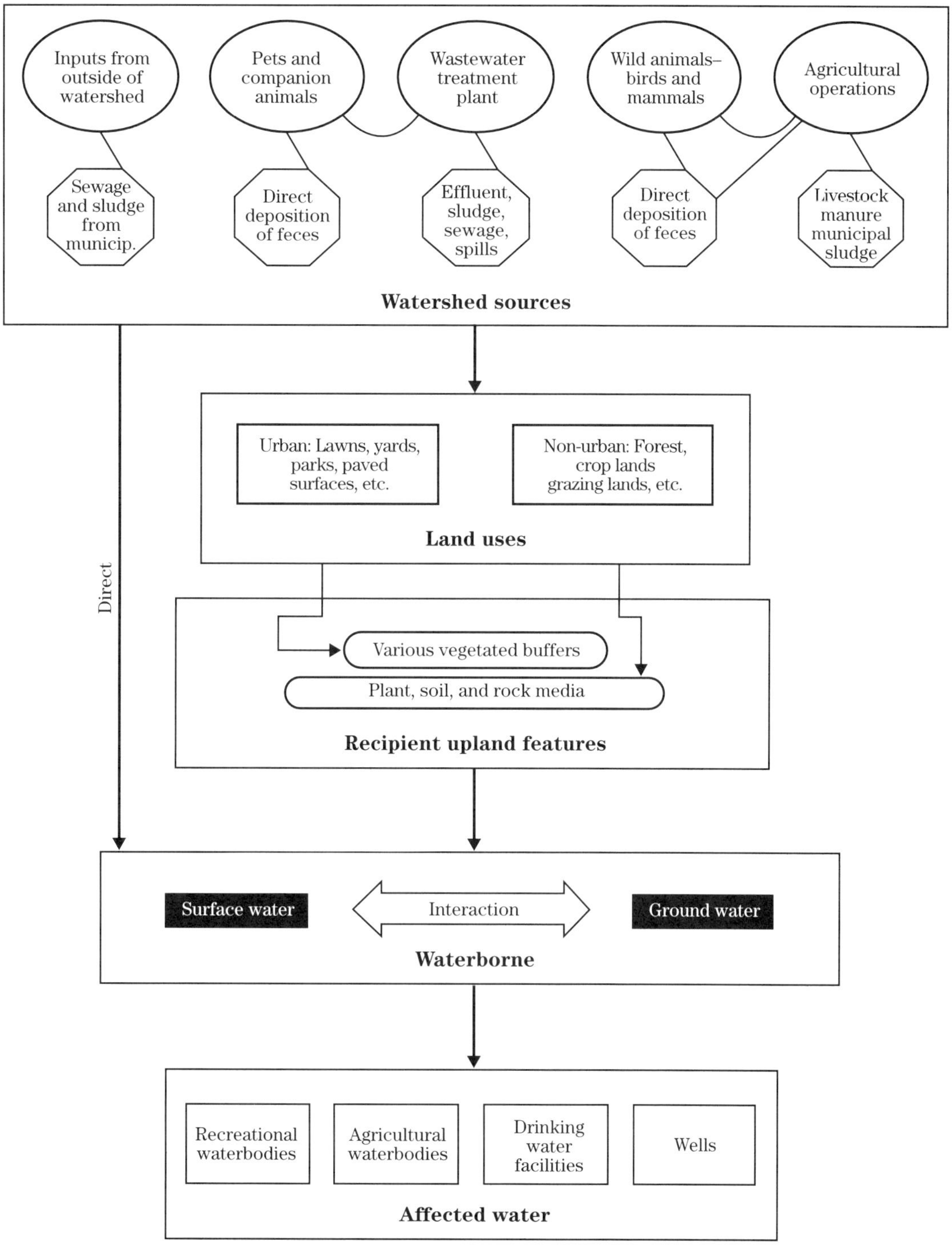

restriction. Class B sludge has received treatment to ensure that pathogens are not likely to pose a threat to public health or the environment, to reduce, but not eliminate pathogens. When farmers use Class B sludge, they are advised to avoid direct human contact or inhalation of dust or spray during and after application. The EPA should be consulted for a detailed review of regulations and technologies of pathogens in biosolids (USEPA, 1992).

Many pathogens can survive sewage treatment, although their numbers are reduced (CAST, 1992; Strauch, 1991). Some pathogens are adsorbed to particles that remain with the sludge during sedimentation processes. A number of cases have occurred where animals have become infected from the spreading of sewage biosolids that were not disinfected. The salmonellosis in dairy cows from pastures in Switzerland is an example (Strauch, 1991). The effluent from the sewage treatment process is generally disinfected by chlorination or other processes and either directly discharged to surface water or applied to land areas. Certain organisms, such as *Cryptosporidium*, are known to survive this disinfection process. **For this reason, uses of water downstream of sewage discharges must be carefully controlled and the water treated appropriately to avoid infection by such organisms.**

Some generalities can be made about survival of pathogens that may be useful in the implementation of conservation practices. For example, table 5 shows survival times of pathogens in freshwater, sewage, on crops, and in soils. Septic tank effluent may be the most significant source of pathogenic bacteria and viruses in the subsurface environment (CAST, 1992). In the mid 1980s, the overflow or seepage of sewage primarily from septic tanks and cesspools was responsible for 43 percent of the reported outbreaks and 63 percent of the reported cases of illness caused by the use of untreated ground water (Craun, 1985).

Table 5 Survival times of pathogens in various media [1/]

Micro-organism	Survival (days)				
	slurry	fecal paddies	soil	water	plants
Bacteria					
Salmonella sp.	250+	200+	150+	16	
E. coli	300+	200+	200+	35	
General maxima [2/]			1 year		180
Viruses					
General maxima			1 year		60
Protozoan cysts					
General maxima				180+	5 (dry)
Helminth ova					
General maxima			7 years	—	150 (dry)

1/ EPA, 1992; Kelly, 1978.
2/ General maxima is the longest time that the organism can survive.

Pets and companion animals

Pets and companion animals are often overlooked as a source of water contamination. Yet their large numbers and the manner in which their waste is disposed can be of concern. Pets provide a potential reservoir for a number of pathogens including *Giardia* sp., *Cryptosporidium parvum*, and *Salmonella* sp. Dogs and cats release waste in yards and walking areas often adjacent to streams that are subject to direct runoff. Young horses are also potential carriers of *Cryptosporidium parvum* (Coleman, 1989; Johnson et al., 1997) and *Giardia* (Johnson et al., 1997). Salt Lake City has enacted strict regulations for companion animals in the Wasatch and Oquirrh Mountains. These mountains supply drinking water for the city, and coliform counts indicated a potential pathogen problem was developing. Management of companion animal waste is extremely variable by some pet owners, ranging from careful collection to complete neglect.

Wildlife—mammals and birds

Wild mammals and birds also act as pathogen reservoirs. They are dispersed across forest land, idle land, pastureland, cropland, and the urban landscape. Their wastes most commonly enter surface water, although leaching to ground water can occur. Wildlife can contribute pathogenic microbes, such as *Salmonella* sp. and *E. coli* (Moorhead and Davis, 1998). Concentrations of wild animals or waterfowl in watersheds can cause problem levels of indicator bacteria. The high density activities of these animals close to or in water provide little opportunity for terrestrial die-off of organisms during their lifecycle.

Pathogen survival in the environment

Certain pathogens exit the host in an environmentally resistant form (e.g., cyst) that can remain viable in the environment for a while. This allows the organism an opportunity to be ingested by another host and repeat its lifecycle. Some pathogens are shed in great numbers, with the random chance that a few organisms will find a new host to repeat the lifecycle. In either case, the survival of the pathogen outside its host is dependent on many variables, including temperature, soil type, and moisture content. Each pathogen needs to be examined individually for a complete understanding. The following introduces the survival of pathogens in an agricultural setting.

Enteric organisms differ widely in their ability to survive various environments outside the host. Once outside the host, the organisms are subjected to stresses and processes (figure 8, page 26).

Waste management systems

The floors of an animal facility are a major collection point for bacteria and parasites that leave their animal host. Most facilities have means of managing wastes from these areas. Manure is stored at various consistencies that are a function of the total solids. The total solid content of manure depends on the animal species, housing, type of bedding used, and addition of other waste and water. Generally, the manure is considered to be in a solid form when its solids comprise at least 20 percent of its mass (see Agricultural Waste Management Field Handbook, figure 9–1). A slurry is defined as a mixture of feces and urine, which may contain cleaning water, rainwater, and small amounts of feed and bedding.

Figure 8 Factors affecting the viability of pathogens along transport pathways

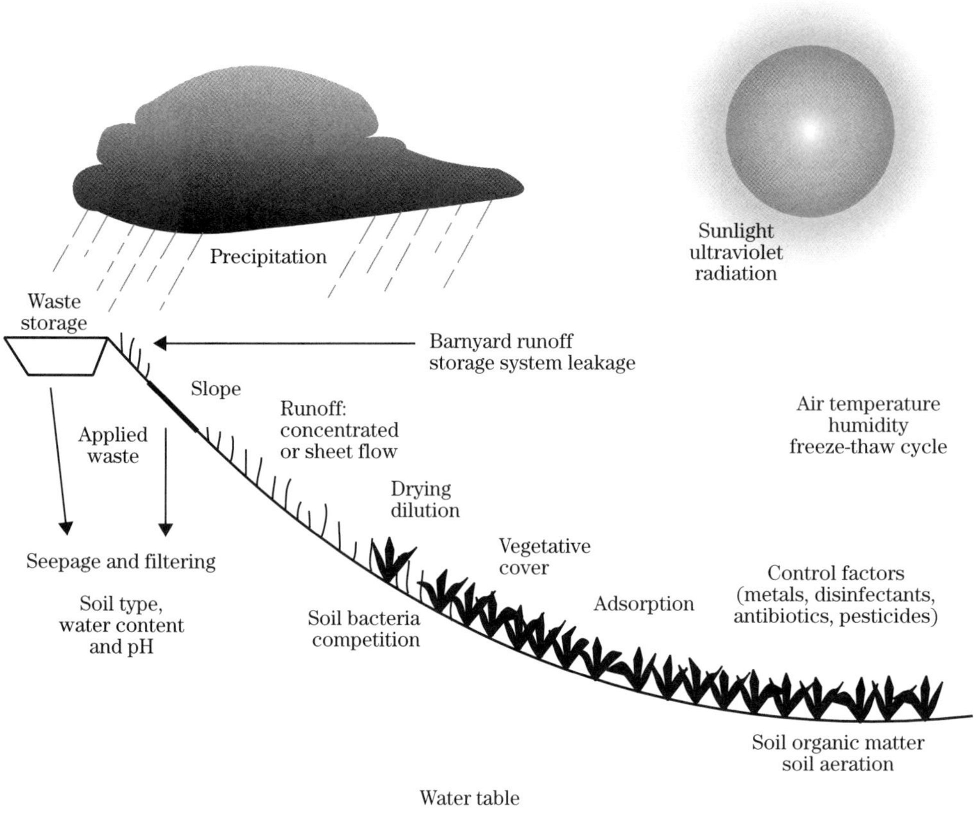

Although reports have shown *Salmonella* reproduction in slurry manure storage, most pathogens decline in solid manure storage (Jones, 1976; Jones, 1979). Solids present the greatest opportunity for spontaneous generation of heat in storage through composting. The resulting high temperature may be sufficient to kill part of the microbial population (figure 9). Some pathogens are destroyed by spontaneous generation of heat in well-bedded manure packs, a technique for disinfection referred to as dung packing. A Swedish study using infected cattle examined both slurry and manure. *Salmonella* was in 35 percent of the slurry samples and only in 6 percent of manure with straw bedding samples for similar herds (Strauch, 1991).

The survival of pathogens in manure, particularly in anaerobic manure, is of primary concern. Organisms are known to survive longer in the anaerobic state than in aeration. This is most likely because the generation of heat from bacterial breakdown of organic material in aerated material is sufficient to shorten bacteria life spans. In a review of several pathogenic species of bacteria (*Salmonella typhimurium*, *E. coli*, *Pseudomonas aeruginosa*, *Staphalococcus aureus*) in cattle manure, Strauch (1987) illustrated that organisms survived up to 15 days in aerated liquid manure, but up to 39 days in nonaerated liquid manure. The inside temperature of the aerated manure ranged from 29 to 34 degrees Celsius (84–93 °F), while that of the nonaerated ranged from 9 to 23 degrees Celsius (48–73 °F). Other environmental factors remained nearly the same for both treatments.

Table 6 on page 28 lists factors that affect the survival of pathogenic micro-organisms during animal waste slurry storage.

Some waste management systems include biogas generation as a component. Thermophilic anaerobic digestion at 49 to 54 degrees Celsius (120–130 °F) can reduce the number of pathogens. Slurry in mesophilic biogas generators only attains a temperature of 30 to 38 degrees Celsius (86–100 °F) and will not destroy as many pathogens. Effluents from mesophilic biogas generators should be considered as having the same hygienic status as any slurry under normal storage conditions to assure that species tolerant of higher temperatures are destroyed (Strauch and Ballarini, 1994).

Figure 9 Time and temperature requirements for 100 percent die-off on selected pathogens from sewage sludge and septic tanks

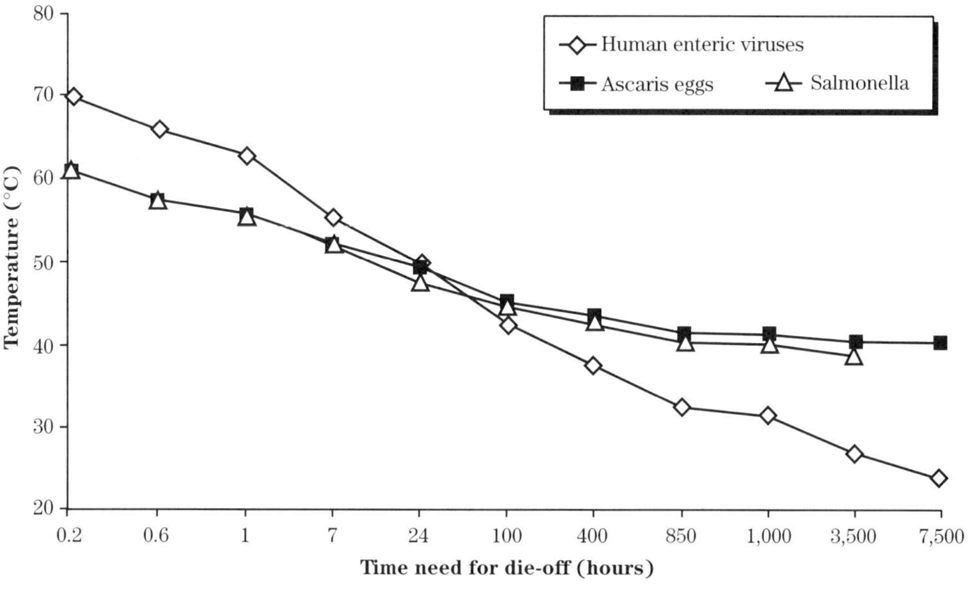

Table 6 Factors affecting survival of pathogens during the storage of animal waste as slurry [1/]

Factor	Considerations
Size of loading dose	Percent of herd infected (higher potential for shedding) Frequency of waste additions (continued supply of organisms) Amount of dilution (lower concentration, less reproductive success for bacteria)
Length of storage period	Die-off rate
Treatment	Anaerobic, aerobic, aeration and composting processes (die-off rate dependent on temperature — duration) Disinfectants Metals and/or pesticide content
Incidental treatment in storage	Aerobic digestion Anaerobic digestion Drying (die-off from desiccation) Freeze/thaw cycle
Slurry characteristics	Available oxygen (for aeration) pH Temperature (see figure 9, page 27) Nutrients (available for sustenance, particularly organic forms) Total solids content (important to some organisms) Inhibitors (antibiotics may have minimal effect — organism resistance)
Innate survival of organism	Vegetative form vs. spore, endospore, or cyst (high resistance to external environment)

1/ Adapted from Kelly, 1978.

Land application areas

Vegetation—Manure applied to pasture falls either on the forage or to the soil surface. Some *Salmonella* species in droppings from infected animals grazing in pastures have been known to survive for more than 28 weeks (Jones, 1979). Some *Salmonella* sp. in digested sewage sludge spread on grass has been recovered viable up to 5 weeks (Strauch, 1991). Survival partly depends on the height of grass. For example, *Salmonella dublin* survives for 10 days on the upper part of the grass and from 14 to 19 days at the lower part of the stem. This difference in survival rates is most likely a result of moisture retention (humidity) and shading (Taylor and Burrows, 1971).

Salmonella longevity also varies by plant species. Those species that have stems and leaves providing refuge or shading extend the viability of bacteria. Visible light, ultraviolet radiation, and desiccation (drying) are the probable factors in destroying the organisms on forage. Rainfall can extend their life by providing needed moisture and washing the organisms from the plant into the soil environment. Hancock et al. (1997b) surveyed 36 dairy herds in Idaho, Oregon, and Washington and found that manure application to forage crops was not associated with the prevalence of *E. coli* O157:H7 in cattle on farms.

Soil—Manure and sewage sludge can be applied by incorporation or injection into cropland soil. Once in the soil subsurface, two major factors control microbial fate: survival and time of migration. The longer a micro-organism persists through its soil migration, the greater the chance it may infect a potential host once it reaches the ground water. Other than helminths, the survival of various micro-organisms has been reported to vary from less than 30 days to more than a year in soil. *Salmonella* after being applied to soil has survived for more than 20 weeks (Findlay, 1972). Stoddard et al. (1998) found nondetectable levels of fecal bacteria in leachate 60 days after application of manure.

Pathogen survival in soil is limited by several factors. Entrapment of micro-organisms by the soil can lead to mortality that is hastened by environmental factors, such as dryness, pH, predator soil micro-organisms, lack of percolation water, and lack of organic matter. Figure 10 illustrates the effect of various soil types on the recovery of *E. coli* at various depths, demonstrating the greatest potential movement of pathogens to ground water in sandy soil compared to clay soil.

Figure 10 Effect of selected soil texture types (without preferential flow) and depth on *E. coli* numbers in effluent

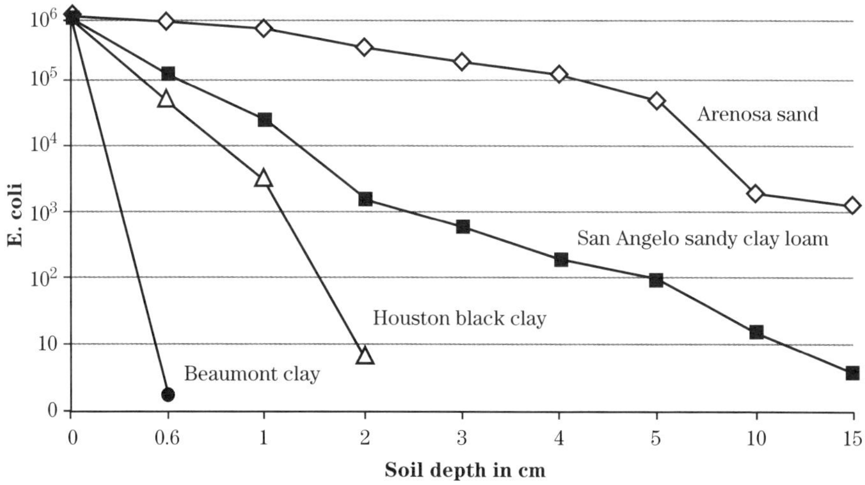

Adapted from Moore et al., 1998

Organic and clay particles in soil effectively trap viruses and smaller bacteria and protozoa (Mawdsley et al., 1995). This occurs as microbes are adsorbed to clay and some negatively charged surfaces. Larger bacteria, protozoa, and helminths are filtered by narrow pore size and through bridging of individuals at pore constrictions. In coarser textured soil, human viruses are more likely to migrate through pores.

The microbial population of soil is restricted to the aqueous phase and the solid-liquid interface. Movement of bacteria and viruses increases in saturated soils. Percolating water provides a mechanism for downward movement. Saturated water flows through macropores, bypasses the filtering effect, and transports micro-organisms long distances.

E. coli cells were found to move as much as 1,500 centimeters per hour through saturated hillside slopes (Rahe et al., 1978). Plant roots tend to increase the translocation of bacteria through soil (Mawdsley et al., 1995). Percolating water can accelerate this movement through these root channels. Soil structure can also provide macropores, as frequently seen in clay soils.

Survival in water

Once a pathogen leaves the host environment, it must adjust to external conditions that are different and often stressful. Biological, chemical, and physical stressors interact in complex ways that have not been well studied. The survival of most pathogens, once discharged into a waterbody, is highly variable depending upon the quality of the receiving water, particularly turbidity, oxygen levels, presence of pesticides and nutrients, temperature, pH, organic matter content, and solar radiation (Moore et al., 1988). Once in a waterbody, micro-organisms often become adsorbed to organic matter and soil particles. These settle out and accumulate at the bottom of rivers and lakes, and may become a source of organism if resuspended. Many small protozoa feed on bacteria, including pathogens, and many invertebrates feed on both bacteria and protozoa in a waterbody.

Extensive research on coliforms, such as *E. coli*, provides the basis for understanding factors affecting bacterial survival in streams, rivers, and lakes. Pathogenic bacteria generally are not well suited to aquatic systems, and survival studies indicate that native bacterial flora outcompete them for nutrients (Korhonen and Martikainen, 1991). In laboratory studies of lake water, *E. coli* can survive for prolonged periods (as long as 260 days) when kept at 4 degrees Celsius (39 °F) (Flint, 1987) when no other bacteria are present. Low temperatures, which slow metabolism, generally prolong the survival of pathogens.

Most bacterial pathogens are sensitive to temperatures exceeding 60 degrees Celsius (140 °F). Bacterial pathogens, which produce resistant endospores or have thick-walled cells, may survive higher temperatures and only be killed by prolonged heating at temperatures in excess of 100 degrees Celsius (212 °F) (Kelly, 1978). Higher temperatures also kill protozoan cysts. Freeze-thaw cycles, when the organisms are subjected to repeated freezing and thawing, also causes pathogen mortality.

The normal pH range for most waterbodies is close to 7 (neutral) and would not affect bacterial survival. Only at extreme pH (<4.5 and >8.2) can cell die-off be anticipated.

Nutrient enrichment of water may play an important role in survival. Lim and Flint (1989) demonstrated the importance of nitrogen in the survival of *E. coli.* They allowed cells to survive the competition from the indigenous bacterial flora. The additional nitrogen appears to be important in allowing cells to go through a period of progressive cell dormancy that prolongs their viability.

The effect of ultraviolet radiation on bacterial and protozoan mortality has long been known. Recent evidence shows visible light as another inhibitor of cell survival (Barcina et al., 1990). *E. coli* and *Enterococcus faecalis* were significantly reduced when exposed to visible light in both freshwater and marine systems.

Estimating enteric bacteria die-off—Moore et al. (1988) provided a way of estimating die-off rates for bacteria. Several algorithms presented are essentially variations of Chick's Law, a simulation of a simple first-order reaction in chemical kinetics given as:

$$\frac{N_t}{N_0} = 10^{-kt}$$

where:

N_t = number of bacteria at time t
N_0 = number of bacteria at time 0
t = time in days
k = first order or die-off rate constant

Although several variations of this equation have been used to more closely portray organism survival under varying environmental conditions, Chick's Law continues to provide a simple, easy to use approach for estimating die-off. Example die-off rate constants, k, for various conditions are provided in table 7 on page 32. For more detail consult Moore et al. (1988).

Table 7 Die-off rate constants for various wastes and conditions [1/]

Material/study	Organism	pH	Season or temperature (°C)	Type of study	Die-off rate, k (days^{-1})
Dairy manure					
Pile	Fecal coliform	---	Oct–Feb	Field	0.066
Pile – covered	Fecal coliform	---	Oct–Feb	Field	0.028
Stored slurry	*Salmonella dublin*	---	Feb	Lab	0.107–0.428
(anaerobic/ inoculated)	*E. coli*				0.102–0.287
Swine manure	*E. coli*	7.0	4	Lab	0.686
(inoculated)		8.0	4		0.867
(slurry storage)		9.0	4		0.931
		7.0	20		0.588
		8.0	20		0.816
		9.0	20		1.079
Swine manure	Fecal coliform	6.4	0–25	Soil: fsl	0.47
applied to grass			0–25		
field plots			0–25		
Poultry manure					
applied to bare	Fecal coliform	4.5–6.5	25	Soil: cl	0.342
soil plots	Fecal strep				0.093
Stream water	*E. coli*	8.4	4–6	Field	1.970
		8.1	4–6		3.140
	E. coli	8.1	5	Lab	0.151
			10		0.231
			15		0.495
			20		0.990

1/ Moore et al. (1988).

Measures for the control of pathogens from agricultural sources

It is beyond the focus of this publication to describe all the potential points of control for waterborne pathogens. Herd health issues, including the clinical diseases associated with pathogens, require the expertise of a veterinarian or an animal healthcare provider. This section focuses on the points of control that have an application in agricultural settings.

A multiple-barrier approach can help control pathogen transport and proliferation. Four control points are illustrated in figure 11 on page 34. They are:

- Pathogen import to the farm, which is intended to prevent the initial infection by these organisms
- The cycle of pathogen amplification or proliferation in the animal operation
- Waste management
- Pathogen export or transport from the farm

These control points should not be treated separately. For example, waste management is an important part of the amplification/proliferation control point when feed becomes contaminated with waste. Waste management also is an important part of the export control point; adequate treatment, such as composting, may have killed the pathogens before they leave the farm.

Several conservation practices have a role in reducing pathogen load in a watershed (table 8). Details on each conservation practice are available from the NRCS National Handbook of Conservation Practices (NRCS, 1999) [**http://www.ncg.nrcs.usda.gov/nhcp_2.html**]. From a watershed perspective, any practice that reduces runoff and erosion will reduce the transport of pathogen directly to surface water.

Table 8 Multiple-barrier approach and selected conservation practices that can reduce pathogen loading to watersheds

Practice 1/	Import control (source)	Amplification/ proliferation	Waste management	Export control (transport)
Composting facility (317)		x	xxx	x
Constructed wetland (656)			x	xxx
Filter Strip (393) and grassed waterway (412)				xxx
Residue management (329 & 344)			x	xxx
Riparian buffers (390 & 391)				xxx
Nutrient management (590)			xxx	
Sediment basin (350)				xxx
Waste management system (312) [including: Waste storage facility (313) Waste treatment lagoon (359) Waste utilization (633)]		xxx	xxx	
Irrigation water management (449)				xxx
Prescribed grazing including use exclusion (472) and Fence (382)			xxx	xxx

1/ Example practices selected from the NRCS National Handbook of Conservation Practices (number indicated-NRCS, 1999). A single **x** indicates an indirect relation to pathogen control, while **xxx** indicates a direct relationship between a practice and pathogen control. Additional practices may also be appropriate.

Figure 11 Multiple-barrier points of pathogen control in agricultural operations

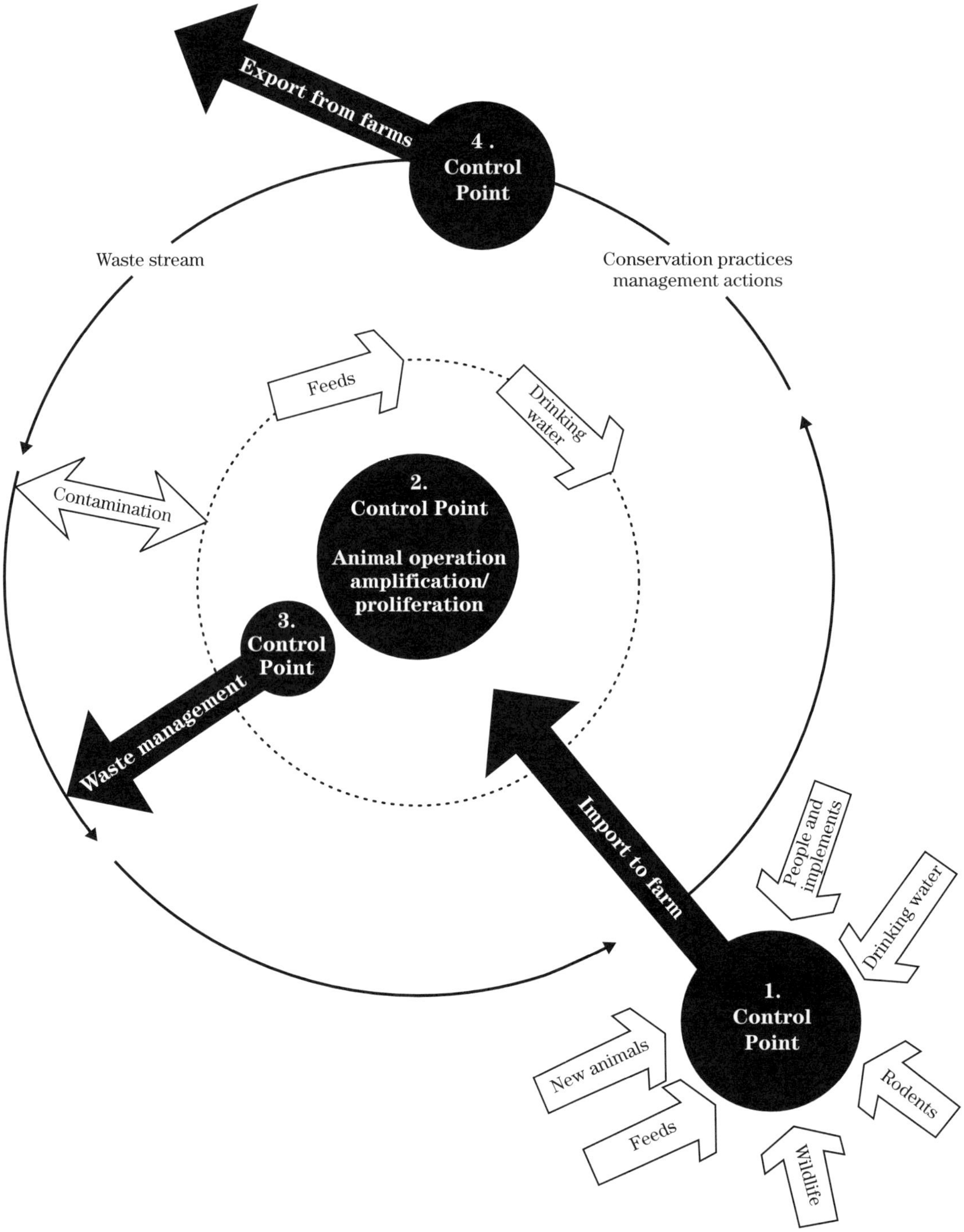

Control point: Import to farms

Feeds

A primary avenue for pathogens to access animals is through the feeding process. Animal feed brought into the farm needs to be free of pathogens, especially for the susceptible, young animals. Wild mice have been demonstrated to have a high prevalence of *Cryptosporidium* and *Salmonella* infection to calves (Klesius et al., 1986), which can easily contaminate grain. Rodents are also a source of *Salmonella* contamination of feed on farms.

Salmonella contamination of feed is recognized by the Food and Drug Administration, and is addressed in their Analysis Critical Control Point Program for Food Safety (FDA, 1994). Any feed contaminated with rodent feces should be suspect, and feed imports as well as onfarm handling of feeds are important.

Some farms purchase hay and green fodder from outside the farm, or they rent pasture. These feed sources can become contaminated from the spreading or deposition of wastes. Appropriate waste handling practices need to be followed before this forage is used. Areas receiving waste applications should not be used for at least 30 days following waste application (based on *Salmonella* - Strauch and Ballarini, 1994; Jones, 1979).

Litter from poultry has been investigated as cattle feed from the perspective of pathogens. Dried and composted poultry litter was not a significant source of *Salmonella*, *Campylobacter* or *E. coli* O157:H7 (Jeffrey et al., 1998).

Drinking water

Contaminated drinking water may be a source of pathogens. Keeping livestock and manure out of drinking water sources, such as streams and ponds, reduces pathogen contamination by direct deposit. Prescribed grazing, which includes exclusion fencing and alternative watering facilities (spring development, troughs, water rams, solar and cattle operated pumps), is an important measure to consider. Onfarm handling of the waste stream must prevent waste from accidentally mixing with a clean drinking water source. Surface water sources can also be contaminated by improper practices in the watershed. Contamination of ground water that comes into a farm and then is used as drinking water is less frequent. Monitoring of this water source for pathogens should be considered if unexplained disease outbreaks occur.

New livestock

When an animal operation acquires new stock from outside the farm, separation or quarantine of the new animals is a practice that may prevent pathogen transmission to healthy animals. The veterinarian or animal health care provider can be consulted for additional measures that may decrease the risk of pathogen introduction from new animals. Some animals that are carriers of pathogens are of special concern because they shed pathogens, yet do not develop illness.

People and implements

Anyone who visits the farm may be a carrier of pathogenic organisms. This is a special concern for those who visit or make deliveries to multiple farms on a regular basis. Farm-to-farm transmission can be prevented by common sense biosecurity measures, such as thorough cleaning of boots, gloves, and any animal handling equipment including vehicle tires.

Control point: Amplification and proliferation

A major goal of any animal operation is the prevention of pathogen spreading within the farm. Animal health is an essential part of that control and may include immunization, treatment, culling, and other measures needed to maintain animal health and reduce the growth and spread (amplification) of an infection on the farm.

Several common sense practices reduce amplification of infection cycles in an animal operation. One major source of pathogens is feed and drinking water contamination by manure in the barn or feeding areas. This point of control is critical, and appropriate actions include:

- frequent cleaning of housing units
- use of separate implements for feeding and cleaning, including shovels, brooms, or hoses
- frequent cleaning of equipment, such as tractor tires and blades
- prevention of foot traffic contamination of feed with manure
- prevention of contamination with fecal material from other animals, such as other agricultural animals, humans, birds, pets, and wildlife
- no feeding of potentially contaminated leftover adult feed to young stock

Waste management

Storage and treatment

Some bacteria are unique in their ability to rapidly proliferate under the right conditions outside the host. Viruses, helminths, and most protozoa, on the other hand, cannot multiply outside their specific hosts. Once reduced by treatment or attrition, their populations stay reduced (USEPA, 1992). To reduce the potential for contamination, manure solids known to contain pathogens can be spread in areas of low hydrologic sensitivity. Hydrologically sensitive areas are those areas were water moves. These areas are time-sensitive. For example, during the dry season, some parts of a flood plain may not be considered hydrologically sensitive.

Composting temperatures for solid manure is known to eliminate the targeted organisms and is another control method.

Several options are available for treatment of manure transferred from animal operations. These options include aerobic lagoons, anaerobic lagoons, controlled anaerobic digestion for methane, composting, and constructed wetlands.

Lagoons are used to reduce pollution potential of the waste through biological, physical, and chemical processes. *Composting* solid manure, if properly managed, provides an opportunity to attain elevated temperatures needed to destroy various organisms including pathogens and plant seeds. *Constructed wetlands* are used to process effluent that, for various reasons, cannot be land applied. All of these components reduce pathogen populations; some more effectively than others.

Anaerobic lagoons reduce the organic content of wastes primarily in the absence of oxygen. Farm lagoons are generally low rate systems and receive no supplemental heat. Thus, temperatures in these lagoons normally do not elevate to more than a few degrees above ambient. Studies on municipal sludge processed anaerobically indicated this process may reduce viruses and coliform bacteria to a greater extent compared to aerobic sludge (Farrah et. al, 1981).

Aerobic lagoons reduce the organic concentration of manure where minimizing odors is critical. Temperatures in these lagoons also remain within a few degrees above ambient. In each of these methods, pathogens are reduced by being deprived of organic material that is needed for survival, and also as a result of the length of storage.

Composting is an effective means of reducing pathogen concentrations. Temperatures over 55 degrees Celsius (131 °F) are easily attained and maintained for sufficient duration. All but the most persistent organisms (e.g., certain viruses and worm eggs) can survive. The main concern in composting is to assure that the entire waste mass is uniformly treated and there are no hot or cold spots. Therefore, a high degree of management is required to completely mix the composted material.

Constructed wetlands can be an important tool in the management of liquid animal waste, including swine, dairy and poultry (CH2M Hill and Payne, 1997). Constructed wetlands can have an important role in confined animal operations, providing a cost-effective treatment system that improves water quality.

Constructed wetlands effectively reduce bacterial pathogen numbers (Watson et al., 1989). Monitoring of constructed wetlands at several small municipal systems in various climates showed removal efficiencies from 82 to 100 percent for fecal coliforms (Hammer, 1989). Gravelbed constructed wetlands have been demonstrated to decrease fecal coliform by 90 to 99.9 percent (Tanner et al., 1998). Constructed wetlands remove organisms as follows:

- Viruses may be adsorbed by soil and organic particles.
- Bacteria are removed by sedimentation, ultraviolet radiation, chemical reactions, natural die-off, and predation by zooplankton.

The control of *Cryptosporidium* and *Giardia* through constructed wetlands is unknown. In some instances, the wildlife that wetlands attract may actually cause a net increase in pathogens.

Application

The main strategy for managing the waste stream following storage and/or treatment is to control the release of pathogens from the waste application areas. Climatic conditions, application technique, timing, and where applications are made are the major considerations for minimizing the loss of micro-organisms in runoff and leaching. Agricultural Waste Management Field Handbook (NRCS, 1992), chapter 5, table 5–3 should be consulted for a detailed review of restricting features that need to be considered during waste application.

Application techniques

On cropland, two approaches to application of animal waste should be considered depending on the field conditions that will most likely follow application.

The first approach, incorporating the waste into the soil, is used when storm events are anticipated. This allows the reduction of potentially harmful organisms through adsorption, filtration, and attack from predator organisms. Direct incorporation also reduces the potential for surface applied waste to be carried away by surface runoff. Incorporating surface-applied waste and injection of waste are effective practices to move organisms into the soil profile. In a study of sod-covered versus tilled soil, McMurry et al. (1998) found greater infiltration of fecal bacteria in sod-covered soil blocks compared to the tilled soils. The soil was well-structured. They also concluded that infiltration in these soils contributed to fecal coliform getting into shallow ground water. Irrigation soon after manure application moved fecal coliform into the drainage effluent in less than an hour (Goehring et al., 1999) and should be avoided.

The second approach to waste application, surface application without incorporation, can be taken when the soil is dry and at summer temperature. This approach allows significant pathogen die-off (Vendrell et al., 1997) by exposure to UV light and desiccation.

Timing

Timing of waste applications is also important especially on established grasslands where incorporation into the soil is not feasible. If waste is applied at the beginning of a dry weather period, the number of organisms can be reduced. On the other hand, waste applied just before a rainfall period will most likely move into runoff. An exception may be in karst soil where there is little runoff. Spring no-tillage manure application on this soil showed no greater fecal bacteria in leachate compared to fall applications (Stoddard et al., 1998).

If waste reaches a waterbody in spring or summer, bacterial growth is stimulated by moderate water temperatures. Proliferation may be the most important factor responsible for the observed seasonal increase in indicator bacteria (Edwards et al., 1997).

If the grassland area is to be grazed, waste that has been stored at least 60 days should be applied at least 30 days before the scheduled grazing period. This may need to be modified if there is evidence to support a longer wait (Strauch, 1991). Use of these areas for grazing should be limited to mature animals.

Application of waste to forage vegetation should take place when the minimal amount of forage biomass is present. This allows sunlight and desiccation to destroy the greatest amount of pathogens and reduces the chances of pathogen adherence to the forage. The minimal forage biomass is present just after harvest or grazing.

Winter application should be avoided in hydrologically sensitive areas, because most pathogens survive longer at cold temperatures, although they do not proliferate, and the potential for runoff is high from frozen or saturated soil.

Hydrologic considerations

Hydrology of the waste application area deserves special attention. Waste placement should be avoided in areas of major flow paths. The major drainage areas of the waste application area can be determined and avoided, when possible.

Although some research indicates that runoff volume does not influence indicator bacterial counts (Edwards et al., 1997), one part of certain watersheds may be the most significant source of surface runoff and its associated pathogen. Watershed runoff processes are affected by soil permeability, soil profile storage, slope steepness, soil complexity, flood plains, frozen soils, surface and subsurface flow paths, antecedent moisture, and saturated conditions.

Saturated areas, also known as critical source areas (Gburek and Pionke, 1995), expand and contract rapidly into adjacent areas during storms in response to precipitation, soils, topography, ground water level, and moisture status. These areas are dominated by saturated overland flow and rapidly respond to subsurface flow. They often form in areas where subsurface lateral flow converges, depth to an impervious layer decreases, or an abrupt change in slope occurs. Thus, a strategy in humid areas may involve identifying the boundaries of critical source areas and applying waste outside those boundaries to minimize pathogen runoff.

Some tools have been or are being developed to delineate critical source areas (e.g., Cornell Soil Moisture Routing Model-Frankenberger, 1999). However, more general delineations can be made using the following factors:

- areas of concentrated flow
- soil drainage class
- flooding frequency
- wetland mapping
- aerial photo interpretation

Control point: Export from farms

The final barrier to pathogen movement into the watershed includes the control of diseased animals or contaminated animal products, grazing land management, and physical strategies, such as buffers and filter strips. The containment of pathogens from contaminated animal products and diseased animals is not addressed in this publication. In practical terms, the nutrient management component of the overall conservation plan includes conservation and erosion prevention measures designed to reduce runoff from areas that have manure applications or intensive animal use. These practices may have a similar effect on reducing the transport of pathogens off the farm and into the watershed. Although conclusive evidence that one specific practice is more effective than another is lacking, general trends are apparent (Meals, 1989). As pathogens move off the farm, any pathogens present will most likely move with them.

Grazing land management

An important control measure is appropriate grazing land management techniques. Along with waste application precautions to grazing lands, these techniques include proper grazing use (wastes distributed more evenly), avoidance of overgrazing to reduce runoff, and keeping animals out of waterbodies to avoid direct deposits of pathogens. For the latter, the most commonly used approaches include alternative watering facilities, salt/molasses, and feed and shade away from open water. Fencing can keep livestock out of streams and direct them to crossings.

Specific watershed management strategies for *Cryptosporidium*

Overall, management strategies designed to minimize direct livestock contamination of surface water with *Cryptosporidium* should focus on the young animals (less than 3 months old). These efforts should focus on cow herds when calves are present (Ongert and Stibbs, 1989; Atwill et al., 1998; Atwill et al., 1999). Handling of calves and their manure is the critical point of control with *Cryptosporidium* and other pathogens. Those operations that normally separate young animals include feedlots, some slaughterhouses, dairies who do not raise their own calves, and backgrounder/stocker beef cattle operations (calves 6 to 14 months old).

Buffers and filter strips

The effectiveness of riparian buffers and filter strips in trapping sediment and associated contaminates is well documented. The primary mechanism of filter strips involves the infiltration of runoff. When the volume of runoff exceeds the infiltration capacity of a riparian buffer, then movement of water and associated pathogens into the receiving waterbody should anticipated.

Although buffers and filter strips can also trap fecal waste (Dickey and Vanderholm, 1981; Doyle et al., 1977; Young et al., 1980), their effectiveness in reducing pathogen transport is not readily apparent (Coyne et al., 1995). Their removal rate is often as high as 95 percent (Coyne and Blevins, 1995); however, the number of indicator bacteria in runoff is initially in the 10^6 cells per 100 milliliter range, leaving about 50,000 cells per 100-milliliter organisms that may still reach the waterbody. This amount of bacteria still greatly exceeds the standards (table 3, page 13).

Flow length is an important consideration. Buffers and filter strips are often wider than practical for most farmers. Coyne and Blevins (1995) used grass filter strips that were 4.5 meters wide and concluded that they deter fecal contamination of water from manured fields on most occasions. Only significant runoff-producing rainfall was a problem. Hubbard et al. (1999) recommend a 10- to 20-meter-wide vegetated filter strip between animal confinement areas and watercourses. They also recommend that wastewater from animal operations not be applied to areas of land that are subject to surface runoff into a watercourse.

Moore et al. (1988) summarized the work of several investigators and suggested that vegetative filters are most reliable in the removal of bacteria at high concentrations (at least 10^5 organisms per 100 milliliters) from waste effluent. The bacterial populations in runoff from these buffer areas seem to equilibrate at about 10^4 to 10^5 organisms per 100 milliliters regardless of the experimental conditions. For this reason, buffers and filter strips should be considered as secondary practices in control and should only be used in conjunction with a full suite of source, proliferation, and treatment control measures (composting, lagoon processing, and constructed wetlands) as a part of the overall waste management system. These buffers become the final polishing effort before runoff entering waterbodies. All feasible practices along the pathway from source to the edge of the farm should be implemented first.

Grass, shrubs, and trees along the riparian interface with cropland and pasture may be effective treatment of overland and shallow subsurface flows for nitrogen and particulate phosphorus removal under the correct conditions. Moore et al. (1988) and Palone and Todd (1997) identified the important mechanisms involved in this reduction as:

- a reduction in the volume of runoff from increased infiltration
- a decrease in runoff velocity caused by the vegetative cover with a resultant increase in sedimentation of pollutants that are adsorbed to particulate matter
- increased adsorption of pollutants by soil particles under a lower ionic concentration regime than found on the waste application site
- dilution by incoming rainfall
- physical adhesion to vegetation and soil particles

Watershed-scale efforts

Watershed-scale research has been conducted to evaluate the collective effectiveness when multiple practices and Best Management Practices (BMPs), including buffers and filter strips, are implemented. The following examples illustrate the potential reduction in fecal bacterial contamination without specifically targeting the results of an individual practice. One study, the St. Albans Bay Rural Clean Water Project (RCWP), showed a fecal bacterial reduction trend when all watershed waste management efforts were collectively examined (figure 12). Monitoring of fecal coliform and fecal streptococcus in the tributaries during the implementation period showed a close association with increasing proportions of watershed animals under BMPs.

In Oregon, the Tillamook Bay RCWP directly addressed water quality impairment resulting from high fecal coliform levels. The problem was caused primarily by manure in runoff from the dairy farms. High fecal coliform levels were causing potential health hazards and negatively affecting commercial oyster industry, recreational clam digging, fishing, boating, and other tourist activities. Many of the oyster beds were closed periodically because of excessive coliform levels. The primary objective of the project was to reduce fecal coliform levels by 70 percent so the deleterious effects being caused to the oyster and tourism industries

Figure 12 Mean annual fecal coliform (FC) and fecal streptococcus (FS) counts in relation to animal units under BMPs for the St. Albans watershed

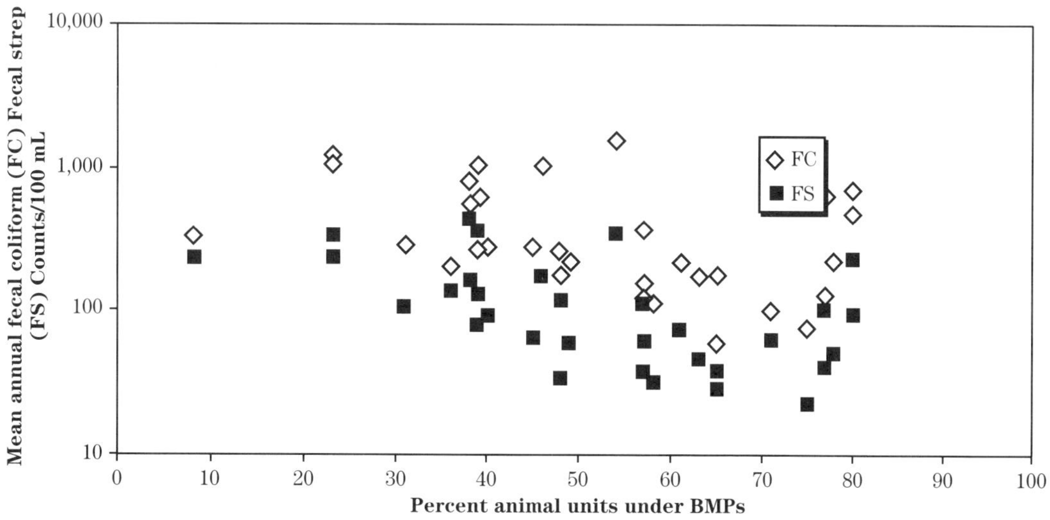

would also be reduced. The influence of BMPs on the water quality of Tillamook Bay is inconclusive at this time (Tillamook Bay Internet communication) although water quality improvements have most likely occurred.

In a study of three Utah watersheds, Glenne (1984) found that percent reduction of bacteria could be correlated to the buffer strip length and slope (Moore et al., 1988). Figure 13 provides a plot of some of these results. It provides only an indication of potential reduction in bacterial numbers in the Utah watersheds where concentrations were high.

In summary, further investigation is needed into the effectiveness of buffers and filter strips for removal of pathogens. It appears they can be effective in removing bacteria from effluent with higher concentrations (at least 10^5 per 100 milliliters), but that reductions below 10^4 per 100 milliliters cannot be expected. Application of percent removal rates to concentrations below this level is misleading. It should be emphasized that buffers should be used as a final polish in the context of an overall, integrated treatment effort on the farm and in the watershed.

Figure 13 Effectiveness of filter strips for bacterial removal (from Utah watershed study, Glenne, 1984)

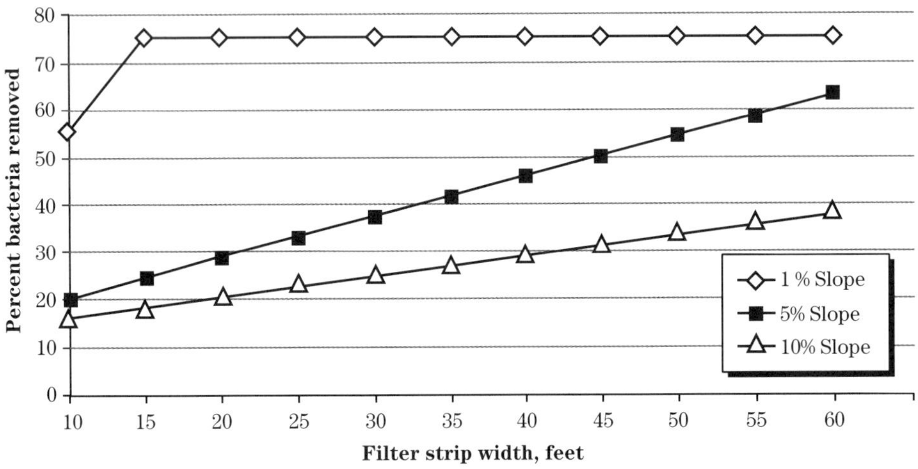

Monitoring and evaluation of pathogens

The NRCS National Handbook of Water Quality Monitoring describes methods for monitoring the water quality response to land use and land management activities and conservation practices. This handbook should be consulted for designing a water quality monitoring project and analyzing the monitoring results.

This technical note is intended to present an overview of selected, specific evaluation methods related to particular pathogens.

Coliform bacteria

Identification and enumeration techniques for coliform bacteria are inexpensive and simple to perform and some have been developed into field kits (Pedley and Bartram, 1996).

Fecal bacterial counts in rivers and lakes around the world that have little human impact vary from less than 1 to as many as 3,000 organisms per 100 milliliters. Remote mountainous areas may have up to 100 organisms per 100 milliliters; however, they generally are below 30 organisms per 100 milliliters. Waterbodies that receive waste from around high population densities can have counts of up to 10 million organisms per 100 milliliters (Chapman, 1996). Ground water not directly connected to surface water should contain no fecal bacteria.

Are coliforms good and accurate indicator organisms?

Coliforms themselves are not pathogenic and only act as an indicator of fecal contamination. They can come from multiple sources including wildlife, livestock, companion animals, sewage, and soil. The presence of *Cryptosporidium* oocysts and *Giardia* cysts is not correlated with fecal coliform counts in water (Fricker and Crabb, 1998). Some data indicate that fecal bacterial counts derived from a water sampling program do not correlate with important pathogen sources, such as diapered recreational users (Keene et al., 1994). An outbreak of the bacterium *E. coli* O157:H7 in Vancouver, Washington, was traced to a recreation lake where the fecal coliform counts indicated it was safe. *E.coli* O157:H7 is not detected by the normal culturing technique used to enumerate *E. coli* in water samples.

Sewage is a major factor in waterborne outbreaks, and coliforms are useful as a first warning. If used properly, a site that has high coliform counts should start the process of a more detailed investigation, usually by public health officials. If the site has unusually high numbers and is used as a recreational site (i.e., swimming), it may be closed immediately.

Fingerprinting or ribotyping (see sidebar on page 44) is a new method that is used to identify the sources of bacterial contamination in some watersheds (Samadpour and Chechowitz, 1995). This method, also called multiple source tracking, is an expensive technique.

Detection methods

Several evaluation methods are used in testing for the presence of enteric bacteria in water and solids. Positive identification for specific organisms may at times be necessary and is often complex for pathogenic bacteria, such as *Salmonella* or *Campylobacter* sp. These organisms usually occur in low numbers in water samples and may need to be concentrated through filtration and enrichment techniques. Two of the more common counting and identification methods use the multiple tube and the membrane filter techniques. In the multiple tube method, a sample is processed through a prescribed series of fermentation tube tests, which involve several sequences of carefully controlled incubations and observations for presence of organisms. A statistical analysis of the results in terms of the number of fermentation tubes processed provides the estimated density of the fecal bacteria expressed as most probable number (MPN) per 100 milliliters. The membrane filter or colony count technique involves filtration of the sample, controlled incubation including the appropriate medium, and a colony count of the bacteria. The density of the organisms is then computed and is normally expressed as number per 100 milliliters. This number is actually in colony forming units (CFU) and may underrepresent the actual individuals because a colony might have originated from a clump of bacteria. Tests for bacteria conducted on semi-solids, such as manure, are normally expressed in number of micro-organisms per unit mass (dry weight basis unless noted otherwise).

A test for the presence of viable helminth ova is available. The test monitors for *Ascaris* ova, which serve as an indicator of several helminth species (USEPA, 1992). Viable helminth ova are observed under a microscope and counted as individuals.

Methodologies for the detection of viruses are constantly being improved, but they require expensive laboratory facilities.

The samples may need to be prepared for transport to highly specialized laboratories. A test is available that simultaneously monitors for several enteric virus species that are presumed to be good indicators for other types of enteric viruses. Viruses are usually counted as plaque forming units (PFU). Each PFU represents an infection zone where a single infectious virus has invaded and infected a layer of animal cells in culture.

Cryptosporidium and *Giardia* were traditionally observed microscopically from fecal material or a concentrated water sample. Recent improvements include the use of fluorescent dyes coupled with antibodies that specifically adhere to these organisms including the ELISA technique. The main difficulty is determining if the observed organisms are alive or not.

Fingerprinting to identify sources

Two unrelated watersheds in the Mid-Atlantic Region have high fecal coliform levels, and managers have been perplexed as to the source. The Philadelphia Suburban Water Company and Montgomery County, Pennsylvania, were faced with finding more sources to treat so that Deep Creek Lake could be reopened to swimmers. Many of the usually suspected sources had already been treated. A small Chesapeake Bay watershed on Virginia's Eastern Shore was overloading a shellfish area with fecal coliforms and threatened to close down the clam business in the area. Again, it seemed that traditional sources (leaky onsite wastewater systems and farms) were exclusively the source.

Scientists tracked fecal coliform levels in the field to generally identify source areas. In the Deep Creek Lake Watershed, two microbiologists applied different genetic identification techniques, called microbial source tracking (MST), to arrive at similar conclusions: resident geese, ducks, and other wildlife accounted for more than 70 percent of the *E. coli* samples taken in the lake (CTIC, 1997). In the Chesapeake watershed, microbiologists developed a library of more than 200 DNA patterns to which numerous samples were compared. They concluded that deer and raccoon were primary sources (Simmons, 1997).

Watershed evaluation of pathogens

Frequently problems associated with waterborne pathogens involve multiple sources within a watershed, such as wildlife, domestic animals, and humans, so that sources may be difficult to isolate. Monitoring strategies generally include a reach-by-reach analysis of the main river stem and tributary inputs to identify locations of high pathogen concentration in an effort to isolate the source(s). The evaluation may become quite complex to determine the principal sources as in the Deep Creek Lake Watershed, Pennsylvania, or a tributary of southern Chesapeake Bay, Virginia.

In watershed evaluations, monitoring of indicator bacteria provides the baseline data needed to identify segments of water bodies with limited uses. Source(s) may be difficult to isolate using monitoring techniques without the expenditure of considerable resources, so information is gained from the ratio of fecal coliform to fecal strep (see table 3, page 13).

Modeling is an alternative to intensive monitoring. Models are useful tools not only to help identify sources, but to also estimate the relative loading from various management scenarios.

Models of indicator bacteria for surface runoff are available for site and watershed scales (table 9). Generally, the logic in these models is to generate organism numbers at the source, assume there is no regrowth, and using literature or observed coefficients or algorithms for die-off, route the populations through various pathways to the impaired water body or threatened location.

Table 9 Selected models that simulate indicator bacteria

Model name	Scale	Main land use	Hydrology	Time scale	Main input data	Available from
Simple methods:						
EPA screening procedures	Watershed	Mixed watershed	N/A	Mean annual	Watershed data, land use, loading factors (default values)	NTIS PB86122496 (EPA/600/6-85/002a), (703) 487-4650
Mid-range:						
Auto-QI	Watershed	Urban	Water balance	Storm event, continuous	Hourly, daily rainfall; watershed data, land use, BMP removal rates	IL Water Survey, 2204 Griffith Dr., Champaign, IL 61820
MWASTE/ MSPRED	Farm	Feedlot/ field/ pasture	ARS-CREAMS	Storm event, annual	Animal waste system; hydrology; field and buffer strip parameters; coliform conc. in runoff from various waste management systems	Environmental engineer, National Water & Climate Ctr, USDA-NRCS, 101 SW Main St., Ste 1600, Portland , OR 97204-3224

Table 9 Selected models that simulate indicator bacteria—Continued

Model name	Scale	Main land use	Hydrology	Time scale	Main input data	Available from
SLAMM	Watershed	Urban	Small storm-based coefficient	Storm event, continuous	Hourly rainfall, characteristics of pollution sources, areas, soil type, imperviousness traffic structure	Dr. Robert Pitt, Dept. Civil/Environmental Engineering, UAL, 1150 Tenth Ave. S Rm 257, Birmingham, AL 35204-4401 (205) 934-8430
Detailed:						
BASINS	Watershed	Mixed land uses	HSPF—water balance of land surface and soil processes	Storm event, continuous	Meteorologic & hydrologic, land use distribution & characteristics, loading factors & washoff parameters, receiving water characteristics, decay coefficients, GIS & national data bases	<http://www.epa.gov/ost/BASINS/> BASINS 3.0 scheduled for release in fall 2000 will incorporate ArcView-SWAT with the capability to simulate bacteria; current version <http://www.brc.tamus.edu/swat/>
SWMM	Watershed	Urban	Nonlinear reservoir	Storm event, continuous	Meterologic & hydrologic data, land use distribution, decay rates; accumulation & washoff parameters	<http://www.ccee.orst.edu/swmm/>

Ground water

Ground water can also be affected by pathogens in watersheds in which the soils are permeable, prone to macropore development, or they are underlain by fractured bedrock. The potential movement of pathogens in ground water including viruses, bacteria, *Cryptosporidium* and *Giardia* was reviewed by Robertson and Edberg (1997). Fecal bacteria and other organisms can contaminate springs and wells where waste is spread on permeable soil, where waste is not properly contained in storage, or where animals are concentrated near the water supply. The potential for contamination of ground water from fecal organisms is normally less than that of surface water because of the soil's filtering action and the antagonistic competition with other soil organisms. Short circuits to natural filtering, such as uncapped or improperly capped wells, are a major source of fecal bacterial contamination in deeper ground water systems. Because most individual ground water supplies are not treated and have zero tolerance (EPA acceptable limit) as it relates to potable water, protection of ground water supplies is a critical consideration.

Anticipated developments from ongoing pathogen research

Detection of protozoan cysts in environmental samples

Cysts in water and soil are hard to detect. Membrane filters have been introduced that may require less water. An immunofluorescent antibody test, which stains cysts after they have been recovered from the filters, helps detect the protozoan cysts. Research is currently in progress to produce more efficient and effective methods for isolating cysts from water. Methods proposed to date increase efficiency in sampling or sample processing and analysis by using filters or other analytical methods (Krir and Genthe, 1993; Nieminski et al., 1995; Fricker and Crabb, 1998). The EPA recently released a methodology (Method 1623) for detecting *Giardia* cysts and *Cryptosporidium* oocysts (USEPA, 1999c). Easily used tests are needed to determine the species and host of origin for *Giardia* cysts and *Cryptosporidium* oocysts, especially in environmental samples (water and soil). These tests would make it possible to determine whether cysts are of public health concern; i.e., infective to humans.

Risk analysis

As part of its role of ensuring that major regulations proposed by USDA are based on sound scientific and economic analysis, the Office of Risk Assessment and Cost Benefit Analysis (ORACBA) is undertaking a series of case studies to apply a multiple criteria decisionmaking technique called Adaptive Risk Analysis (ARA) to USDA's Resource Conservation Programs (Sadeghi et al., 1998; Tumeo et al., 1998 — Internet communications). In the case studies, ecological evaluation is an essential part of the overall ARA method. Ecological endpoints, such as nutrients from fertilizers and sediment from soil erosion, have been estimated using environmental fate models. However, with the increased use of manure in various agricultural production systems and heightened concern over potential ecological and human health impacts, the fate and transport of micro-organisms resulting from manure application need to be better understood.

Initial findings

Microbial scenarios for various pathogens associated with different types of manure in terms of their persistence, die-off, and regrowth rates have been completed and are incorporated into the SWAT (Soil and Water Assessment Tool) model. Work is underway to use several existing microbial transport data bases on the fate and transport of pathogens from manure for initial testing and validation of the microbial component of the model. However, a more precise calibration and validation of the model is expected to be done after more quantitative time zero microbial population and transport data are obtained from the controlled field lysimeters. These lysimeters are designed specifically to evaluate the transport capability of *Cryptosporidium* oocysts in overland waterflow conditions.

Application of results

The incorporation of the newly developed SWAT model in the ARA method provides a more comprehensive tool to examine the success of USDA's Environmental Quality Incentives Program (EQIP). This newly developed process-oriented tool allows a look at specific watersheds and determines the impact of a range of management practices, such as manure management strategies, on multiple environmental and human health objectives. This information provides the basis for comparison of risk to the costs associated with a set of potential alternative management strategies.

Pathogen transport

Conceptual modeling

The transport of *Cryptosporidium* at the watershed level is also being modeled in Canada (Park and Huck, 1999). They have developed a framework for estimating the relative contributions of cattle farms and wastewater treatment plants to *Cryptosporidium* oocysts in watersheds (Zhang et al., 1999). The relationship between waterborne parasites (*Cryptosporidium* and *Giardia*) and beef production in the North Saskatchewan River is also being examined (Cooke, 1999 — Internet communications).

Distribution, dissemination, and fate of *Cryptosporidium* oocysts

An ARS-designed rainfall simulator has been setting the stage for what will probably be the largest outdoor experimental study of the movement of coliform bacteria and *Cryptosporidium parvum* in rain runoff and soil water. The simulator has a boom that flexes as the land's slope changes, keeping the rain the same distance from the soil up and down the slope. The simulator will help scientists trace water movement to track possible routes for bacterial pathogens and for *Cryptosporidium parvum* (Shelton, 1999).

Studies are underway to identify flushwater systems that minimize the risk of environmental contamination and inter-cattle transmission of *Cryptosporidium* and *Giardia* and at the same time conserve water usage through flushwater recycling (Atwill, 1999-Internet communications). Other studies have shown or are being conducted to examine the prevalence of and risk factors for the shedding of *Cryptosporidium* and *Giardia* in livestock populations, llamas, wild pigs, and horses (Atwill et al., 1997; Cole et al., 1998; Wade et al., 1999; Mohammed et al., 1999).

Summary

The vast majority of waterborne pathogens affecting people do not originate from agricultural sources. Human contamination or inadequacies at water treatment plants have been implicated in almost all large-scale waterborne outbreaks. The potential does exist for contamination of water with pathogens from agriculture, warranting a proactive approach for reducing this source in watersheds. Because of the large amount of fecal material produced and used on the farm, as well as the spreading of human waste (sewage and sludge), this potential exists. At this point in time, scientific literature indicates that sound agricultural practices currently in place significantly reduce the opportunity for the introduction of pathogens to the watershed.

This publication introduces the subject of pathogenic organisms and measures used to reduce their transport into the environment. A multiple-barrier approach is recommended. This approach includes prevention of pathogen imports to the farm, breaking the amplification cycle on the farm, proper handling of the waste stream, and control of transport from the farm. On-farm measures need to include a veterinarian or other health care provider who can target control points for pathogen introduction, amplification, and export from the farm before they get into the environment. Although pathogen-specific practices have not been developed, those practices used for waste management and to control runoff help control pathogen transport from the farm to the watershed.

Glossary

Aerobic	A habitat or organism that relies on the presence of oxygen for continued growth and existence.
Amnesic shellfish poisoning (ASP)	An impairment or lack of memory caused by the ingestion of shellfish contaminated with the diatom *Pseudo-nitzschia* sp. that has produced the toxin domoic acid. ASP can be a life-threatening syndrome.
Amoeba	A single-celled (protozoan) organism that constantly changes shape.
Anaerobe	Used to describe a biological habitat or an organism that exists and grows with oxygen.
Asymptomatic	Carrying a particular disease, but not showing any symptoms.
Autoinfection	When a parasite reproduces in a host that then re-infects itself.
Biofilm	A thin layer of organisms that colonize and coat the surface in a water environment.
Campylobacteriosis	An infection of the intestines caused by bacteria of the *Campylobacter* genus. Symptoms of campylobacteriosis include mild to severe diarrhea (often bloody), stomach pain, fever, nausea, and vomiting.
Capsid	A protein coat that covers the nucleoprotein core or nucleic acid of a virus.
Ciguatara fish poisoning (CFP)	Gastrointestinal, neurological, and cardiovascular symptoms associated the ingestion of fish contaminated with toxic dinoflagellates (any of the following: *Gambierdiscus toxicus*, *Prorocentrum* sp., *Ostreopsis* sp., *Coolia monotis*, *Thecadinium* sp. and *Amphidinium carterae*). Paralysis and death have been documented, but symptoms are usually less severe although debilitating.
Ciliate(d)	A grouping of microscopic single-celled organisms that have hair-like projections on their surface used for locomotion.
Coccus	A bacterial cell that has the shape of a sphere.
Coliphage	A bacterial virus (phage) that uses coliform bacteria as a host.
Cryptosporidiosis	A diarrheal illness of varying severity caused by a microscopic intestinal parasitic protozoan, *Cryptosporidium*. It is a common cause of diarrhea worldwide and is common in AIDS patients.
Cyst	A resting stage of an organism that has a tough outer coating.
Dinoflagellate	Photosynthetic organisms of the order Dinoflagellida (for botanists Dinophyceae). They are aquatic and have two flagella that allow the organism to move.
DNA (deoxyribonucleic acid)	The genetic material of all cells and many viruses.

E. coli — Short for *Escherichia coli*, the colon bacillus, a bacterium that normally resides in the intestine of many animals.

ELISA (enzyme linked immuno sorbent assay) — An assay designed for detection of many different organisms, such as certain *Salmonella* sp., *E. coli*, and *Cryptosporidium* (many organism-specific test kits available) http://www.bioline.dk/index.html, http://www.maff.gov.uk/animalh/bse/animal-health/feedban-elisatest.html.

Encyst — The process a cell undergoes to produce a cyst.

Endospore — An asexual spore formed within a bacterial cell.

Enteric — Of or relating to the small intestine.

Epidemiology — The study of populations to determine the frequency and distribution of disease and to measure risks.

Facultative anaerobe — An organism that normally lives in an oxygen-free environment, but can survive oxygen conditions.

Fecal coliform — A grouping of coliform bacteria that lives in the intestines of warm-blooded animals. Elevated measurements of these bacteria in surface water may indicate the presence of human and/or animal waste. Health advisories may be posted when measurements indicate an increased risk to humans from exposure.

Flagellum — A long, thin, hair-like projection from a cell used in movement.

Gastroenteritis — Inflammation of the mucous membrane of the stomach and intestine.

Harmful algal blooms — Elevated growth of one or more species of algae, which may result from excessive nutrient loading in combination with adequate light, temperature, and other environmental factors.

Immunofluorescent antibody — A technique developed to specifically identify an organism. Initially, a probe is developed from the surface properties (antigens) of an organism. These are used to stimulate the production of an antibody that is specific to that species. The antibody is conjugated (bonded) with a fluorescent dye that allows it to be seen.

Inoculated — The process of adding the material containing seed organisms to start growth.

Isolate(s) — An organism that is isolated from a single source, usually by culturing.

Lysogenic phase — The ability of some bacterial viruses to survive in a bacterium as a result of the integration of their DNA into the host chromosome.

Lytic phase — The normal cycle of infection of a cell by a virus in which a mature virus is produced and the cell is then ruptured.

Mesophilic — Requiring a warm temperature in which to develop.

Nonculturable — Not being able to grow an organism in laboratory culture on media.

Oocyst — The environmentally resistant stages of protozoan, such as *Cryptosporidium*.

Organelle — A structurally discrete component of a cell.

Paralytic shellfish poisoning (PSP) — A life-threatening syndrome caused by consumption of contaminated shellfish. The causative organisms are the dinoflagellates: *Alexandrium* sp., *Gymnodinium catenatum*, *Pyrodinium bahamense* that can be accumulated by shellfish. The toxins produced are called saxitoxins.

Parasite — An organism that lives in or on an animal or plant, called a host, and takes its nourishment to the host's detriment.

Pathogen — An agent of disease; it can be a parasite.

RiboNucleic acid (RNA) — A chemical similar to DNA. The several classes of RNA molecules play important roles in protein synthesis and other cell activities.

Ribotype — The RNA signature or fingerprint of a cell used to identify the source of the organism.

Rotavirus — A virus that is the leading cause of severe winter diarrhea in young children.

Salmonellosis — An infection of the intestines caused by the *Salmonella* bacteria, which causes severe diarrhea and death in some cases.

Serotype — A technique used to determine different stains of bacteria.

Shedding — The releasing of organisms from the host, usually in feces.

Shigella — A group of bacteria that normally inhabits the intestinal tract and causes infantile gastroenteritis, summer diarrhea of childhood, and various forms of dysentery.

Spirillum — A fairly rigid, helically twisted bacterial cell often, but not necessarily, a member of the genus *Spirillum*.

Spirochete — Bacteria that appear worm-like, spiral-shaped, and wiggle vigorously when viewed under a microscope.

Sporozoites — A stage in the development oocyst of *Cryptosporidium* that infects intestinal cells.

Thermophilic — Requiring high temperature in which to develop.

Trophozoite — The feeding stage of a protozoan (as distinct from reproductive or encysted stages).

Zoonotic — A disease caused by pathogens that are transmitted among animals and humans.

Zooplankton — The floating, often microscopic animals that live in aquatic environments.

References

Acha, P.N., and B. Szyfres. 1987. Zoonoses and communicable diseases common to man and animals. Washington, DC: Pan American Health Organization.

Adak, G.K., J.M. Cowden, and S. Nicholas. 1995. The public health laboratory service national case-control study of primary indigenous sporadic cases of *Campylobacter* infection. Epidemiol. Infect. 115:15–22.

Altekruse S.F., N.J. Stern, P.I. Fields, and D.L. Swerdlow. 1999. *Campylobacter jejuni*-An emerging foodborne pathogen. Emerging Infectious Diseases 5:1–10.

Altekruse, S.F., J.M. Hunt, and L.K. Tollefson. 1994. Food and animal sources of human *Campylobacter jejuni* infection. J. Am. Vet. Med. Assoc. 204:57–61.

Anderson, B.C. 1986. Effect of drying on the infectivity of cryptosporidia-laden calf feces for 3- to 7-day old mice. Amer. J. Vet. Research 47:2272–2273.

Atwill, E.R. 1996. Assessing the link between rangeland cattle and water-borne *Cryptosporidium parvum* infection in humans. Rangelands 18:48–51.

Atwill, E.R., E. Johnson, D.J. Klingborg,. G.M. Veserat, G. Markegard, W.A. Jensen, D.W. Pratt, R.E. Delmas, H.A. George, and L.C Forero. 1999. Age, geographic, and temporal distribution of fecal shedding of *Cryptosporidium parvum* oocysts in cow-calf herds. Amer. J. Vet. Research 60:420–425.

Atwill, E.R., J.A. Harp, T. Jones, P.W. Jardon, S. Checel, and M. Zylstra. 1998. Evaluation of periparturient dairy cows and contact surfaces as a reservoir of *Cryptosporidium parvum* for calfhood infection. *Amer. J. Vet. Research* 59:1116–1121.

Atwill, E.R., R.A Sweitzer, M.G.C. das Pereira, I.A. Gardner, D. VanVuren, and W.M. Boyce. 1997. Prevalence of and associated risk factors for shedding *Cryptosporidium parvum* oocysts and *Giardia* cysts within feral pig populations in California. Appl. Environ. Microbiol. 63:3946–3949.

Atwill, R. 1999. (Internet communications) Protozoal parasites, livestock, and water quality. University of California, Davis, Veterinary Medicine Extension, Veterinary Medicine Teaching and Research Center (VMTRC), Tulare, CA. http://www.vetnet.ucdavis.edu/vetext/16-WaterQuality.html

AWWARF. 1999. Workshop on Emerging Pathogens web site: http://www.awwarf.com/newprojects/empath.html.

Badenoch, J. 1990. *Cryptosporidium* in water supplies. Report of the Group of Experts. HMSO, London, 230 pp.

Barcina, I.J., M. Gonzalez, J. Iriberri, and L Egea. 1990. Survival strategies of *Escherichia coli* and *Enterococcus faecalis* in illuminated fresh and marine systems. J. Appl. Bacter. 68:189–198.

Besser, T.E., D.D. Hancock, L.C. Pritchett, E.M. McRae, D.H. Rice, and P.I. Tarr. 1997. Duration of detection of fecal excretion of *Escherichia coli* O157:H7 in cattle. J. Infec. Dis. 175:726–9.

Black, R.E., M.D. Levine, and M.L. Clements. 1988. Experimental *Campylobacter jejuni* infection in humans. J. Infect. Dis. 157:472–479.

Blair, K. Internet communications. Cryptosporidium and Public Health. http://www.c3.org/library/dwhn010395.html.

Blewett, D.A. 1989. Quantitative techniques in *Cryptosporidium* research, 85–95. *In* K.W. Angus and D.A. Blewett (eds.), Proc. 1st Int. Workshop.

Bolton F.J., D. Coates, D.N. Hutchinson, and A.F. Godfree. 1987. A study of thermophilic campylobacters in a river system. J. Appl. Bact. 62:167–76.

Bowman, D.D., M.B. Jenkings, and W.C. Ghiorse. (in progress). Distribution, dissemination and fate of *Cryptosporidium parvum* oocysts. Department of Microbiology and Immunology, Cornell University, Ithaca, NY.

Buret, A., N. denHollander, P.M. Wallis, D. Befus, and M.E. Olson. 1990. Zoonotic potential of giardiasis in domestic ruminants. J. Infect. Dis. 162:231–237.

Burton, G.A., D. Gunnison, and G.R. Lanza. 1987. Survival of pathogenic bacteria in various freshwater sediments. Appl. Envir. Microbiol. 53:633–638.

Carter A.M., R.E. Pacha, G.W. Clark, and E.A. Williams. 1987. Seasonal occurrence of *Campylobacter* sp. and their correlation with standard indicator bacteria. Appl. Environ. Microbiol. 53:523–6.

CH2M Hill. 1997. Constructed wetlands and wastewater management for confined feeding operations. Prepared for EPA's Gulf of Mexico Program. Available by calling this program at (228)-688-1033.

CH2M Hill and Payne Engineering. 1997. Constructed wetlands for animal waste treatment: A manual on performance, design, and operations with case studies. Prepared for EPA's Gulf of Mexico Program. Available by calling this program at (228)-688-1033.

Chapman, D. (ed.). 1996. Water quality assessments. second edition. E & F.N. Spon–Chapman and Hall. NY.

Chomel, B.B., E.E. DeBess, and D.M. Mangiamele. 1994. Changing trends in the epidemiology of human brucellosis in California from 1973–1992: a shift toward foodborne transmission. J. Infect. Dis. 170:1216–1223.

Cole, D.J., N.D. Cohen, K. Snowden, and R. Smith. 1998. Prevalence of and risk factors for fecal shedding of *Cryptosporidium parvum* oocyts in horses. J. Am. Vet. Med. Assoc. 213:1296–1302.

Coleman, S.U. 1989. Prevalence of *Cryptosporidium* sp. in equids in Louisiana. Amer. J. Vet. Research 50:575–577.

Conservation Technology Information Center. 1997. DNA whodunit. Vol. 15(6):15–16.

Cooke, S. (1999–In progress) Relationship between waterborne parasites *(Cryptosporidium* and *Giardia)* and Beef Production in the North Saskatchewan River http://www.fbminet.ca/manure/97AB080.htm.

Council for Agricultural Science and Technology (CAST). 1992. Water quality: agriculture's role, task force report 120, Ames, IA.

Coyne, M.S. R.A. Gilfillen, R.W. Rhodes, and R.L. Blevins. 1995. Soil and fecal coliform trapping by grass filter strips during simulated rain. J. Soil and Water Conser. 50:405–408.

Coyne, M.S., and R.L. Blevins. 1995. Fecal bacteria in surface runoff from poultry-manured fields. *In* K. Steele (ed.), Animal Water and the Land-water Interface, Lewis Publishers, CRC Press, Boca Raton, FL, pp. 77–87.

Craun, G.F. 1990. Waterborne Giardiasis. *In* E.A. Meyer (ed.), *Giardiasis*, Vol. 3, Series in Human Parasitic Diseases, E.J. Ruttenberg and A.J. MacInnis (eds.). New York: Elsevier, pp. 267–293.

Cray, W.C. Jr., T.A. Casey, and M.A. Rasmussen. 1995. Effect of dietary stress in ruminants on fecal shedding of coliforms and *E. coli* O157:H7. Annual Meeting American Society Microbiology, P7, 383.

deRegnier, D.P., L. Cole, and D.G. Schupp. 1989. Viability of *Giardia* cysts suspended in lake, river, and tap water. Appl. Envir. Microbiol. 55:1223–1229.

Deshpande, P.D., and U.V. Shastri. 1981. Incidence of *Giardia* infection in calves in Maharashtra State, India. Trop. Anim. Hlth. Prod. 13:34.

Dickey, E.C., and D.H. Vanderholm. 1981. Vegetative filter treatment of livestock feedlot runoff. J. Environ. Quality 10:279–84.

Doyle, R.C., G.C. Stanton, and D.C. Wolf. 1977. Effectiveness of forest and grass buffer strips in improving water quality of manure polluted runoff. ASAE Paper 77–2501.

Draser, B.S., and M.J. Hill. 1974. Human intestinal flora: the distribution of bacterial flora in the intestine. London, Academic Press. pp. 36–43.

DuPont, H.L., C.L. Chappell, and C.R. Sterling. 1995. The infectivity of *Cryptosporidium parvum* in healthy volunteers. N. Engl. J. Med. 332:855–859.

Edwards, D.R., M.S. Coyne, P.F. Vendrell, T.C. Daniel, P.A. Moore, and J.F. Murdoch. 1997. Fecal coliform and *Streptococcus* concentrations in runoff from grazed pastures in northwest Arkansas. J. Amer. Water Resources Assoc. 33:413–422.

Erlandsen, S.L. 1994. Biotic transmission-is *Giardiasis* a zoonosis. *In* R.C.A. Thompson, J.A. Reynoldson, and A.J. Lymbery (eds.), Giardia: From Molecules to Disease. Wallingford, UK: CAB International, pp. 83–97.

Erlandsen, S.L., and E.A. Meyer (eds.). 1984. *Giardia* and *Giardiasis* - biology, pathogenesis, and epidemiology. Plenum Press, New York, 400 pp.

Farrah, S.R., G. Bitton, E.M. Hoffmann, O. Lani, O.C. Pancorbo, M.C. Lutrick, and J.E. Betrand. 1981. Survival of enteroviruses and coliform bacteria in a sludge lagoon. Appl. Environ. Microbiol. 41:459–465.

Fayer, R. 1994. Effect of high temperature on infectivity of *Cryptosporidium parvum* oocysts in water. Appl. Environ. Microbiol. 60:2732–35.

Fayer, R., and B.L.P. Ungar. 1986. *Cryptosporidium* sp. and cryptosporidiosis. Microbiol. Rev. 50:458–483.

Fayer, R., and T. Nerad. 1996. Effects of low temperatures on viability of *Cryptosporidium parvum* oocysts. Appl. Environ. Microbiol. 62:1431–33. May not use this if the material does not have normal temperature in it.

Fayer, R.. 1997. *Cryptosporidium* and *Cryptosporidiosis.* CRC Press, New York, pp. 251.

Feng, P., and S.D. Weagant. 1994. Yersinia. *In* Y.H. Hui, J.R. Gorham, K.D. Cliver, and D.O. Cliver (eds.), Foodborne Disease Handbook, Vol 1. New York: Marcel Dekker, Inc. pp. 427–460.

Filice, F.P. 1952. Studies on the cytology and life history of a *Giardia* from the laboratory rat. Univ. CA Publ. in Zoology 57:53–145.

Findlay, C.R. 1972. The persistence of *Salmonella dublin* in slurry in tanks and in pasture. Vet. Record. 91:233–235.

Flint, K.P. 1987. The long-term survival of *Escherichia coli* in river water. J. Appl. Bacter. 63:261–270.

Food and Drug Administration. 1994. Guideline for the utility studies for anti-salmonella chemical food additives in animal feeds. Guideline 54. Docke 94D–0147. (www.fda.gov/cvm/fda/TOCs/guideline54.html).

Franco, D.A., and C.E. Williams. 1994. *Campylobacter jejuni. In* Y.H. Hui, J.R. Gorham, K.D. Cliver, and D.O. Cliver (eds.), Foodborne Disease Handbook, Vol 1. New York: Marcel Dekker, Inc. , pp. 71–96

Frankenberger, J.R., E.S. Brooks, M.T. Walter, and T.S. Steenhuis. 1999. A GIS-based variable source area hydrology model. Hydrologic Processes 13:805–822.

Fricker, C.R., and J.H. Crabb. 1998. Waterborne cryptosporidiosis: detection methods and treatment options. Advances in Parasitology 40:241–278

Gburek, W.J., and H.B. Pionke. 1995. Management strategies for land-based disposal of animal wastes: Hydrologic implications. *In* Steele, K. (ed.), Animal Water and the Land-water Interface, Lewis Publishers, CRC Press, Boca Raton, FL, pp. 313–323.

Geohring, L.D., P.E. Wright, T.S. Steenhuis, and M.F. Walter. 1999. Fecal coliforms in tile drainage effluent. ASAE Paper No. 99-2203, St. Joseph, MI.

Ghiorse, W.C., D.D. Bowman, M.F. Frongillo, M.B. Jenkins, and M.J. Walker. 1997. Parasite viability research support. *In* Science for the Whole Farm Planning-Cornell University Phase II Tenth Quarter Report, for period ending March 31, 1997. New York State Water Resources Institute, Cornell University, Ithaca, NY.

Glenne, B. 1984. Simulation of water pollution generation and abatement on suburban watersheds. Water Resources Bulletin 20:2.

Goldstein, S.T., D.D. Juranek, O. Ravenholt, A.W. Hightower, D.G. Martin, J.L. Mesnick, S.D. Griffiths, A.J. Bryant, R.R. Reich, and B.L. Herwaldt. 1996. Cryptosporidiosis: An outbreak associated with drinking water despite state-of-the-art water treatment. Annals of Internal Med. 124:459-468.

Goodgame, R.W. 1996. Understanding intestinal spore-forming protozoa: Cryptosporidia, Microsporidia, Isospora, and Cyclospora. Ann. Intern. Med. 124:429–441.

Goodgame, R.W., R.M. Genta, and A.C. White. 1993. Intensity of infection in AIDS-associated cryptosporidiosis. J. Inf. Dis. 167:704–709.

Graeber, I., M.A. Montenegro, C. Bunge, and U. Boettcher. 1995. Molecular marker analysis of *Salmonella typhimurium* from surface water, humans, and animals. Europ. J. Epidemiol. 11:325–331

Hammer, D.A. (ed.). 1989. Constructed wetlands for wastewater treatment: municipal, industrial and agricultural. Lewis Publishers, Chelsea, MI.

Hancock, D.D., D.H Rice, D.E. Herriott, T.E. Besser, E.D. Ebel, and L.V. Carpenter. 1997b. Effects of farm manure-handling practices on *Escherichia coli* O157 prevalence in cattle. J. Food Protection 60:363–366.

Hancock, D.D., D.H. Rice, L.A. Thomas, and D.A. Dargatz. 1997a. Epidemiology of *Escherichia coli* O157 in feedlot cattle. J. Food Protection 60:462–465.

Hancock, D.D., T.E. Besser, D.H. Rice, D.E. Herriott, and P.I. Tarr. 1997c. Longitudinal study of *Escherichia coli* O157:H7 in fourteen cattle herds. Epidemiol. Infect. 118:193–5.

Hancock, D.D., T.E. Besser, D.H. Rice, E.D. Ebel, D.E. Herriott, and L.V. Carpenter. 1997d. Multiple sources of *Escherichia coli* O157 in feedlots and dairy farms in the northwestern USA. Prev. Vet. Med. 35:11–19.

Hayes, E.B., T.D. Matte, T.R. O'Brien, T.W. McKinley, G.S. Logsdon, J.B. Rose, B.L.P. Ungar, D.M. Word, P.F. Pinsky, M.L. Cummings, M.A. Wilson, E.A. Long, E.S. Hurwitz, and D.D. Juranek. 1989. Large community outbreak of cryptosporidiosis due to contamination of a filtered public water supply. New England J. Med. 320:1372–1376.

Heath, S.E., and R. Johnson. 1994. Leptospirosis. JAVMA 205:1518–1523.

Hubbard, R.K., J.A. Entry, and J.E. Thies. 1999. Movement of coliform bacteria through riparian buffer systems receiving swine lagoon wastewater. Presented at the July 19–21, 1999 ASAE Annual International meeting, Paper 992100, St. Joseph, MI 49085–9659.

Humphrey, T.J., and P. Beckett. 1987. *Campylobacter jejuni* in dairy cows and raw milk. Epidemiol. Infect. 98:263–9.

Jackson, L.A., A.F. Kaufmann, and W.G. Adams. 1993. Outbreak of leptospirosis associated with swimming. Ped. Inf. Dis. J. 12:48–54.

Jeffrey, J.S., J.H. Kirk, E.R. Atwill, and J.S. Cullor. 1998. Prevalence of selected microbial pathogens in processed poultry waste used as dairy cattle feed. Poultry Science 77:808–811.

Johnson, E., E.R. Atwill, M.E. Filkins, and J. Kalush. 1997. The prevalence of shedding of *Cryptosporidium* and *Giardia* sp. based on a single fecal sample collection from each of 91 horses used for backcountry recreation. J. Vet. Diagnostic Investigation 9:56–60.

Jones, P.W. 1976. The occurrence of salmonellas, mycobacteria and pathogenic strains of *E. coli* in pig slurry. J. Hyg. Camb. 77:43–50.

Jones, P.W. 1979. Health hazards associated with handling of animal wastes. Vet. Record 106:4–7

Juranek, D.D. 1995. Cryptosporidiosis: sources of infection and guidelines for prevention. Clin. Inf. Dis. 21 (Suppl 1): S57–S61.

Keene, W.E., J.T. Barrett, D.W. Fleming, K. Hedberg, F.C. Hoesly, J.M. McAulty, F.L. Oxman, M.A. Pfaller, and L.P. Williams. 1994. Swimming-associated outbreak of hemorrhagic colitis caused by *Escherichia coli* O157:H7 and *Shigella sonnei. New England J. of Med.* 331:579–584.

Kelly, W.R. 1978. Animal and human health hazards associated with the utilization of animal effluents. *In* A Workshop in the EEC Programme of Coordination of Research on Effluents, November 21–23, Dublin, Ireland. Commission of the European Communities. Luxembourg.

Kfir, R., and B. Genthe. 1993. Advantages and disadvantages of the use of immunodetection techniques for the enumeration of micro-organisms and toxins in water. Water Science and Technol. 27:243–252.

Kirkpatrick, C.E. 1985. *Cryptosporidium* infection as a cause of calf diarrhoea. Vet. Clin. North Am. Food Anim. Pract. 1:515:528.

Klesius, P.H., T.B. Haynes, and L.K. Malo. 1986. Infectivity of *Cryptosporidium* sp. isolated from wild mice for calves and mice. JAVMA 189:192–193.

Koenraad, P.M., R. Ayling, and W.C. Hazeleger. 1995. The speciation and subtyping of *Campylobacter* isolates from sewage plants and waste water from a connected poultry abattoir using molecular techniques. Epidemiol. Infect. 115:485–494.

Korhonen, L., and P.J Martikainene. 1991. Survival of *Escherichia coli* and *Campylobacter jejunii* in untreated and filtered lake water. J. Appl. Bacter. 71:379–382.

Kramer, M.H., F.E. Sorhage, S.T. Goldstein, E. Dalley, S.P. Wahlquist, and B.L. Herwaldt. 1998. First reported outbreak in the United States of cryptosporidiosis associated with a recreational lake. Clinical Infectious Diseases 26:27-33.

Landry, M.S., and M.L. Wolfe. 1999. Fecal bacteria contamination of surface waters associated with land application of animal waste. Presented at the July 18–21, 1999 ASAE/CSAE-SCGR Annual International Meeting, Paper No. 994024. St. Joseph, MI 49085–9659.

LeChevallier, M.W., and W.D. Norton. 1995. *Giardia* and *Cryptospori-dium* in raw and finished water. J. Am. Water Works Assoc. 87:54–68.

LeChevallier, M.W., W.D. Norton, and R.G. Lee. 1991. Occurrence of *Giardia* and *Cryptosporidium* sp. in surface water supplies. Appl. Environ. Microbiol. 57:2610–16.

Levine, W.C., W.T. Stepenson, and G.F. Craun. 1991. Waterborne disease outbreaks, 1986–1988. J. Food Prot. 54:71–78.

Levy, D.A., M.S. Bens, G.F. Craun, R.L. Calderon, and B.L. Herwaldt. 1998. Surveillance for Waterborne-Disease Outbreaks—United States, 1995–1996. Morbidity and Mortality Weekly Report 47 (SS–5):1–34.

Lim, C.H., and K.P. Flint. 1989. The effects of nutrients on the survival of *Escherichia coli* in lake water. J. Appl. Bacter. 66:559–569.

Lorenzo, M.J., E. Ares-Mazás, and I. Villacorta Martinez de Maturana. 1993. Detection of oocysts and IgG antibodies to *Cryptosporidium parvum* in asymptomatic adult cattle. Vet. Parasitol. 47: 9–15.

MacKenzie, W.R., N.J. Hoxie, M.E. Proctor, M.S. Gradus, K.A. Blair, D.E. Peterson, J.J. Kazmierczak, D.G. Addiss, K.R. Fox, J.B. Rose, and J.P. Davis. 1994. A massive outbreak in Milwaukee of *Cryptosporidium* infection transmitted through the public water supply. New England J. Med. 331:161-167.

Mawdsley, J.L., R.D. Bardgett, R.J. Merry, B.F. Pain, and M.K. Theodorou. 1995. Pathogens in livestock waste, their potential for movement through soil and environmental pollution. Appl. Soil Ecol. 2:1–15.

Mayo Clinic Health Oasis. 1999. The "bug" that made Milwaukee famous: Trying to tame the tiny parasite *Cryptosporidium*. http://www.mayohealth.org/mayo/9606/htm/ cryptosp.htm.

McMurry, S.W., M.S. Coyne, and E. Perfect. 1998. Fecal coliform transport through intact soil blocks amended with poultry manure. J. Environ. Qual. 27:86–92.

Meals, D.W. 1989. Bacteriological water quality in Vermont agricultural watersheds undergoing land treatment. Lake and Reservoir Management 5:53–62.

Mishu, B., J. Kehler, L.A. Lee, and D. Rodrigue. 1994. Outbreaks of *Salmonella enteritidis* infections in the United States, 1985–1991. J. Infect. Dis. 169:547–552

Mohammed, H.O., S.E. Wade, and S.L. Schaaf. 1999. Risk factors associated with *Cryptosporidium parvum* infection in dairy cattle in southeastern New York State. Vet. Parasit. 83:1–13.

Moore, J.A., J. Smyth, S. Baker, and J.R. Miner. 1988. Evaluating coliform concentrations in runoff from various animal waste management systems. Special Report 817. Agricultural Experimental Stations Oregon State Univ., Corvallis, and U.S. Dep. Agric., Portland, OR.

Moorhead, D.L., W.S. Davis, and C.F. Wolf. 1998. Coliform densities in urban waters of west Texas. J. Environ. Health 60:14–18.

Nieminski, E.C., F.W. Schaefer, III, and J.E. Ongerth. 1995. Comparison of two methods for detection of *Giardia* cysts and *Cryptospori-dium* oocysts in water. Appl. Environ. Microbiol. 61:1714-1719.

O'Handley, R.M., C. Cockwill, T.A. McAllister, M. Jelinski, D.W. Morck, and M.E. Olson. 1999. Duration of naturally acquired giardiosis and cryptosporidiosis in dairy calves and their association with diarrhea. J. Am. Vet. Med. Assoc. 214:391–396.

Olson, M.E., C.L. Thorlakson, L. Deselliers, D.W. Morck, and T.A. McAllister. 1997a. *Giardia* and *Cryptosporidium* in Canadian farm animals. Vet. Parasit. 68:375–381.

Olson, M.E., N.J. Guselle, R.M. O'Handley, M.L. Swift, T.A. McAllister, M.D. Jelinski, and D.W. Morck. 1997b. *Giardia* and *Cryptosporidium* in dairy calves in British Columbia. Can. Vet. J. 38:703–706.

Ongerth, J.E., and H.H. Stibbs. 1989. Prevalence of *Cryptosporidium* infection in dairy calves in western Washington. Amer. J. Vet. Research 50:1069–1070.

Palone, R.S., and A.H. Todd (ed.). 1997. Chesapeake Bay riparian handbook: A guide for establishing and maintaining riparian forest buffers. Chesapeake Bay Program and Northeast Area State and Private Forestry. U.S. Dep. Agric., Forest Service NA-TP-02-97, Radnor, PA.

Park, C.H., and P.M. Huck. 1999. Conceptual modeling of the transport of *Cryptosporidium* oocysts in a watershed. Abstract from the International Symposium on Waterborne Pathogens. Aug 29–Sep 1, 1999, Milwaukee, WI. http://www.awwa.org/99intl/program.htm# monPathogens/infiltra.htm.

Pavlasek, I. 1984. First record of *Giardia* sp. in calves in Czechoslovakia. Folia Parasit. (Praha) 31:225–226.

Pedley S., and J. Bartram. 1996. Microbiological analysis. *In* Bartram and Ballance (ed.), Water Quality Monitoring: A Practical Guide to the Design and Implementation of Freshwater Quality Studies and Monitoring Programmes. E & FN Spon-Chapman and Hall, NY. pp. 237–261.

Pieniazek, N.J., F.J. Bornay-Llinares, S.B. Slemenda, A.J. da Silva, I.N.S. Moura, M.J. Arrowood, O. Ditrich, and D.G. Addiss. 1999. New *Cryptosporidium* genotypes in HIV-infected persons. Emerging Infectious Diseases 5(3). http://www.ama-assn.org/special/hiv/library/readroom/eid99/crypto.htm.

Porter, J.D.H., H.P. Ragazzoni, J.D. Buchanon, H.A. Waskin, D.D. Juranek, and W. Parkin. 1988. *Giardia* transmission in a swimming pool. Amer. J. Public Health 78:659–662.

Quilez, J., C. Sanchez-Acedo, E. del Cacho, A. Clavel, and A.C. Causape. 1996. Prevalence of *Cryptosporidium* and *Giardia* infections in cattle in Aragon (northeastern Spain). Vet. Parasit. 66:139–146.

Rahe, T.M., C. Hgedorn, E.L. McCoy, and G.F. Kling. 1978. Transport of antibiotic-resistant *Escherichia coli* through western Oregon hillslope soils under conditions of saturated flow. J. Environ. Qual. 7:487–494.

Roberts, T. 1989. Human illness costs of foodborne bacteria. Am. J. Agri. Econ. 71:468–474.

Robertson, J.B., and S.C. Edberg. 1997. Natural protection of spring and well drinking water against surface microbial contamination. I. Hydrogeological parameters. Critical Reviews Micro. 23:143–178.

Robertson, L.J., A.T. Campbell, and H.V. Smith. 1992. Survival of *Cryptosporidium parvum* oocysts under various environmental pressures. Appl. Environ. Microbiol. 58:3494–3500.

Robinson, D.A. 1981. Infective dose of *Campylobacter jejuni* in milk. Br. Med. J. 282:1584.

Rose, J.B., C.P. Gerba, and W. Jakubowski. 1991. Survey of potable water supplies for *Cryptosporidium* and *Giardia.* Environ. Sci. Technol. 25:1393–1400.

Sadeghi, A.M., M.A. Tumeo, R. Meekhof, R. Srinivasan, and J.A. Arnold. 1998. Model estimation of pathogenic bacteria transport in overland flow runoff. The Fourth NHEERL Symposium, Extrapolation *In* Human Health And Ecological Risk Assessments. April 27–30, Cary, NC, Abstract, p. 18.

Samadpour, M., and N. Chechowitz. 1995. Little Sooks Creek microbial source tracking. King County Department of Public Works, Surface Water Management Division, and the University of Washington, Seattle, WA.

Sanford, S.E. 1987. Enteric cryptosporidial infection in pigs: 184 cases (1981–1985). JAVMA 190:695–698.

Scott, C.A., H.V. Smith, and M.M.A. Mtambo. 1995. An epidemiologic study of *Cryptosporidium parvum* in two herds of adult beef cattle. Vet. Parasitol 57:277–288.

Shaw, R.D. 1992. Kayaking as a risk factor for leptospirosis. Mo. Med. 89:354–357.

Shelton, D.R. 1999. Tracking manure-borne pathogens. Agricultural Research, November (also http://hydrolab.arsusda.gov/ecl/Pathogens/lysimete.htm).

Shelton, D.R., S.A. Rouhi, and E. Kuczynska. 1999. Internet communications http://hydrolab.arsusda.gov/ecl/Pathogens/infiltra.htm.

Simmons, G., Jr. 1997. DNA fingerprinting aids investigation - fecal coliform sources traced to unlikely suspects. EPA-Nonpoint Source News-Notes, Terrene Institute, Alexandria, VA, 48:19–20.

Smith, H.V., and J.B. Rose. 1990. Waterborne cryptosporidiosis. Parasitol. Today 6:8–12.

St. Albans Bay RCWP 1980–1990. Final Report, May 1991. Vermont RCWP Coordinating Committee.

Stehman, S.M., C. Rossiter, P. McDonough, and S. Wade. 1996. Potential pathogens in manure. *In* Animal Agriculture and the Environment: Nutrients, Pathogens, and Community Relations. Proceedings from the Animal Agriculture and the Environment North American Conference, Rochester, NY, December 11–13, 1996. NRAES–96, Northeast Regional Agricultural Engineering Service, Ithaca, NY.

Steidinger, K.A., J.H. Landsberg, and E.W. Truby. (In prep). *Crypto-peridiniopsis brodyii* gen. et sp. nov. (Dinophyceae): a small lightly armored dinoflagellate similar to *Pfiesteria*.

Sterling, C.R. 1990. Waterborne cryptosporidiosis. *In* J.P. Dubey, C.A. Speer, and R. Fayer (eds.) Cryptosporidiosis of man and animals. Boca Raton, FL, CRC Press. pp. 51–58.

Stern N, D. Jones, I. Wesley, and D. Rollins. 1994. Colonization of chicks by non-culturable *Campylobacter* sp. Letters in Appl. Microbiol. 18:333–6.

Stern, N.J. 1992. Reservoirs for *Campylobacter jejuni* and approaches for intervention in poultry. *In* I. Nachamkin, M.J. Blase, and L.S. Thomplins (eds.), Campylobacter jejuni: Current and Future Trends. Washington, DC, Amer. Soc. Microbiol., pp. 49–60.

Stoddard, C.S., M.S. Coyne, and J.H Grove. 1998. Fecal bacteria survival and infiltration through a shallow agricultural soil: timing and tillage effects. J. Environ. Qual. 7:1516–1523.

Strauch, D. 1987. Hygiene of animal waste management. *In* Animal Production and Environmental Health, Elsevier Science, Amsterdam.

Strauch, D. 1991. Survival of pathogenic microorganisms and parasites in excreta, manure and sewage sludge. Rev. Sci. Tech. Off. Int. Epiz. 10: 813–846.

Strauch, D., and G. Ballarini. 1994. Hygienic aspects of the production and agricultural use of animal wastes. J. Vet. Med. *B*. 41:176–228.

Taminelli, V., and J. Eckert. 1989. Prevalence and geographical distribution of Giardia infections of ruminants in Switzerland. Schweiz Arch. Tierheilk. 131:251–258.

Tanner, C.C., J.P.S. Sukias, and M.P. Upsdell. 1988. Relationship between loading rates and pollutant removal during maturation of gravel-bed constructed wetlands. J. Environ. Qual. 27:448–458.

Tauxe, R.V. 1992. Epidemiology of *Campylobacter jejuni* infections in the United States and other industrialized nations. *In* I. Nachamkin, M.J. Blase, and L.S. Thomplins (eds.), Campylobacter jejuni: Current and Future Trends. Washington, DC, Amer. Soc. Microbiol., pp. 9–19.

Taylor, R.J., and M.R. Burrows. 1971. The survival of *Escherichia coli* and *Salmonella dublin* in slurry on pastures and the infectivity of *S. dublin* for grazing calves. Br. Vet. J. 127:536–543.

Terzieva, S.I., and G.A. McFeters. 1991. Survival and injury of *Escherichia coli*, *Campylobacter jejuni*, and *Yersinia enterocolitica* in stream water. Can. J. Microbiol. 37:785–790.

Thompson, R.C.A., and P.F.L. Boreham. 1994. Discussion report: biotic and abiotic transmission. *In* R.C.A. Thompson, J.A. Reynoldson, and A.J. Lymbery (eds.), Giardia: From Molecules to Disease. Wallingford, UK, CAB International, pp. 131–136.

Tillamook Bay Internet communication. http://h2osparc.wq.ncsu.edu/info/rcwp/orprof.html.

Todd, E.C.D. 1989. Preliminary estimates of costs of foodborne disease in the United States. J. Food Prot. 52:595–601.

Tumeo, M.A., D.A. Mauriello, A.M. Sadeghi, and R. Meekhof. 1998. Case studies on application of adaptive risk analysis to USDA's resource conservation programs. (http://www.usda.gov/oce/oracba).

Tyzzer, E.E. 1910. An extracellular coccidium, *Cryptosporidium muris* (gen. et sp. nov.), of the gastric glands of the common mouse. J. Medical Research 23:487–509.

Tyzzer, E.E. 1912. *Cryptosporidium parvum* (sp. nov.), a coccidium found in the small intestine of the common mouse. Archiv fur Protistenkunde 26:394–412.

United States Department of Agriculture, Natural Resources Conservation Service. 1996. National handbook of water quality monitoring (part 600).

United States Department of Agriculture, Natural Resources Conservation Service. 1992. Agricultural waste management field handbook. National Engineering Handbook, Part 651. http://www.ncg.nrcs.usda.gov/awmfh.html.

United States Department of Agriculture, Natural Resources Conservation Service. 1999. National handbook of conservation practices. http://www.ncg.nrcs.usda.gov/nhcp_2.html.

United States Environmental Protection Agency. 1986. Ambient water quality criteria for bacteria. Report No. EPA440/5-84-002, Washington, DC.

United States Environmental Protection Agency. 1992. Environmental regulations and technology: Control of pathogens and vector attraction in sewage sludge. Report No. EPA/625/R-92/013, Washington, DC.

United States Environmental Protection Agency. 1994. The Quality of our Nation's water: 1994. Office of Water 305 (b) Report, Washington, DC.

United States Environmental Protection Agency. 1996. The Quality of our Nation's water: 1996. Office of Water 305 (b) Report, Washington, DC.

United States Environmental Protection Agency. 1999a. *Pfiesteria piscicida* homepage: http://www.epa.gov/OWOW/ estuaries/ pfiesteria/ July 12, 1999.

United States Environmental Protection Agency. 1999b. U.S. Department of Agriculture U.S. Environmental Protection Agency unified national strategy for animal feeding operations, March 9, 1999, http://www.epa.gov/owm/finafost.htm.

United States Environmental Protection Agency. 1999c. Method 1623: *Cryptosporidium* and *Giardia* in water by filtration/IMS/FA. Report No. EPA-821-R-99-006, Washington, DC, http://www.epa.gov/nerlcwww/1623.pdf.

United States Environmental Protection Agency. 1999d. Action plan for beaches and recreational waters. Report No. EPA-600/R-98/079, Washington, DC.